Hans Keirstead, a global leader of the Stem Cell Revolution and cancer therapy development tells the stories that illustrate the leadership journey of a "Change Agent" with world impact. Vision, commitment, courage – it's all there; read, learn, and be inspired!

Bob Klein | Founder of California Institute of Regenerative Medicine (CIRM), President of Klein Financial Corporation

What I enjoyed the most about *Life, the Universe and Curing Everything* was the blended autobiographical and business success approach. Learning about Dr. Keirstead's life really added greater context to the business and leadership lessons that followed, making them that much more impactful.

Philippe Schaison | President of Allergan (retired)

Two words: Innocent Intrigue. It's captivating advice like this that makes *Life, the Universe, and Curing Everything* a compelling, useful, and fun book to read. Keirstead has really captured the essence of successful leadership and brings it home with his incredible life experiences. If you want to be successful in life and business, read this book!

Neil Sahota | United Nations AI Advisor, Master Inventor of IBM (retired), CEO of ACSI Labs

A wonderful journey through the life and times of one of the foremost medical scientists of our world today! Learning about Dr. Keirstead's pioneering research and how he handled the numerous obstacles in his path was nothing short of inspirational!

Glenn Hopper | CFO, Sandline Discovery

Life, the Universe, and Curing Everything was an absolute page turner! It's funny and sincere at the same time, and the fact that it's told from the perspective of a brilliant stem-cell expert only adds to its already weighty impact.

Shawn Johal | Business Growth Coach, Bestselling Author of The Happy Leader Elevation Leaders

A truly fascinating read! It's not very often we get to peek into the minds of brilliant individuals, but with Hans Keirstead, we most certainly enter that territory. I would highly recommend this book to all business professionals and up-and-coming entrepreneurs!

Rick Yvanovich | CEO, TRG International

In *Life, the Universe, and Curing Everything*, Dr. Keirstead dives deep into the mind of a scientist, the life of a CEO and the facts of curing cancer. Very much an easy read that helps give a bit of entertainment with insight into balancing life while you navigate this crazy world. Remember: Realize that stress will always be an inevitable component of any leadership position, and learn to look on the lighter side of things.

Aaron Vick | Muti-X Founder, Startup Advisor, Author of Leaderpreneur: A Leadership Guide for Startups and Entrepreneurs, and Inevitable Revolutions: Secrets and Strategies for a Successful Business

Witty, inspiring, heartfelt, and genuine... Dr. Keirstead has touched on all the characteristics that go into crafting a great book. His thoughts on leadership and business success are second to none. A great read.

Senator Joe Dunn

Drawing on a wealth of personal and professional experience, Hans Keirstead has presented us with a highly enjoyable collection of some of the biggest life lessons learned from his scientific research, company founding, and from high-level leadership perspectives. Very well done!

Mike Singer | Serial Entrepreneur and Managing Director, Innocreative Capital

Life, the Universe, and Curing Everything

LEADERSHIP PRACTICES FOR CHANGING THE WORLD

HANS KEIRSTEAD

Leaders Press

ISBN 978-1-63735-009-6 (pbk)
ISBN 978-1-63735-008-9 (e-book)

Print Book Distributed by Simon & Schuster
1230 Avenue of the Americas
New York, NY 10020

Library of Congress Control Number: 2021908913

DEDICATION

This book is dedicated to my insanely confident and overreaching father, who always gripped...or at least touched... whatever he reached for.

"Ah, but a man's reach should exceed his grasp, or what's a heaven for?"

—Robert Browning

TABLE OF CONTENTS

INTRODUCTION

In 1855, Henry G. Bohn wrote, "The road to hell is paved with good intentions." I'd like to add that it's also paved with liars, greed, accidents, disease, and cruelty. But I suppose that wasn't politically correct for the times. Probably still isn't.

The point of this little diatribe is to make light of what we already know: very few people walk a straight line to success. Very few people get *it* right— whether that be their lives, careers, or relationship— on their first try. You need balance, you need to put the important things in front, and you need to make serious efforts to ensure you are surrounded by the right people at the right time. In some ways, it's a roll of the universal dice. You close your eyes, jump into the abyss, and hope that if the universe is going to swallow you up, it at least has the decency to do so with a fork and a knife like a civilized person.

I didn't set out to, and certainly could not, write the manifesto on living the perfect life. I've made plenty of mistakes and I'm just as liable to trip over my own two feet as the next person. However, along this crazy rollercoaster ride I've learned a thing or two about people, the universe, and the machinations of life; as a scientist, it sort of comes with the territory.

My aim then in writing this book is to cut away the excess, to give you the most important lessons I've learned as a scientist, entrepreneur, and CEO. I have been extremely fortunate in that my journey has allowed me to cross paths with some brilliant individuals from all over the world whom I am honored to call friends, colleagues, and mentors. Their wisdom, insight and wherewithal are as much a part of my story as my childhood growing up below the poverty line in rural Canada.

But I won't lie to you either—this book has been a challenge.

It's not my natural inclination to posture and strut about like a puffed-up chicken. I'm a big believer in the whole "speak softly but kick ass" thing—well, at least the "speak softly" part. So, why now? Why write this book at all?

It wasn't vanity, and it wasn't to win a popularity contest. The truth is, our world has been irrevocably kicked in the pants by the

COVID-19 virus and it will have far-reaching implications for our global economy. It has given every one of us a greater perspective on the world, and on our place in it. We need a new system; we need a foundation of beliefs and philosophies that will help us cope with the daunting challenges that lie in wait. We need responsible, ethical leaders and business owners. And we need more scientists!

I won't pretend I have all the answers in that regard, but I want to be part of the discussion. If this book can accomplish only one thing, I hope it inspires you—and leads you to manifest that inspiration. I hope it galvanizes your resolve and fortifies your own personal mission. Whether you are a CEO, doctor, lawyer, teacher, student, or parent, there is always someone who is looking up to you.

Let's lead them, together, toward a brighter future.

1—INNOCENT INTRIGUE

Have you ever watched a baby discover her hand? The moment of realization that this thing waving around in her field of vision is actually under her control is a joy to behold. The look of amazement, often followed by one of deep concentration, signals a change of perspective, a widening of her world. At some point, she moves on to experimentation, finding out what can be done with this interesting appendage. It will be some time before she learns this is a hand, but the intrigue of discovery has begun.

EMBRACING THE NOVELTIES OF LIFE

Innocent intrigue, my term for the experience of discovery that shifts our perspective and prompts us to explore, is a core concept in my life. From childhood to at least young adulthood, we have experiences that foster innocent intrigue—a child's first taste of ice cream, a teenager's first time behind the wheel of a car, that newly minted graduate's first day of work. The energy generated by experiences like these fuels our desire to experiment, learn, and experiment some more.

Innocent intrigue generates excitement and pure willpower, unadulterated by preconceptions. Think back to a first-time experience you had early in life. No expectations, no frame of reference, just your senses encountering something brand new. What did that experience create in you? What did you create as a result of the experience?

How about today? When you see something new and awesome, do you have a physical sensation, like tingling in the middle of your chest or a sudden sense of power in your gaze? Are your senses on overdrive? Those are signs you have been hooked, and if you can stay with those sensations and ride the wave of innocent intrigue, you could do something awesome.

As we get older, preconceptions can slow us down considerably and thwart the experience of innocent intrigue. I've observed that the sparks that accompany innocent intrigue are realized a lot less frequently in the careers of many—too many—of my colleagues. Why does this happen?

The human species is predisposed to adapt. It's our superpower. Our senses are poorer than those of most other animals on the planet, our muscles weaker, our very lives more fragile. But we're better at one thing, and that is our ability to adapt. Humans are the best adapters on the planet.

In previous eras, our ancestors routinely faced new experiences, learned from them, and adapted. As human culture has advanced, we have become shielded from many experiences that would normally require us to adapt. Our modern society allows people to shape their environment to be comfortable, and often people don't want to experience any more perspective shifts for fear of losing that comfort. Perhaps accumulated knowledge and experiences are enough for some people to accomplish what they've set their sights on. Comfort, sufficiency, inertia—for whatever reasons, these things decrease innocent intrigue or make it disappear completely as the years go by, if it's not consciously provoked.

I saw this happen among my academic colleagues. When I was young, I was one of 2,000 applicants for a professorship at the University of California. I was good, and so was everyone else. Really good. We were all full of innocent intrigue, ready to tap into the energy and excitement our state of mind naturally generated.

I was selected for the position and eventually joined all the new professors for an orientation session about our new jobs. Looking around this room full of rock stars, I was amazed and honored. Young, hungry, eager, brilliant people, all of us, and we had competed with literally thousands of others for our jobs. Here we were, at the beginning of a pathway that was funded and fully supported, and our job was to do precisely the work we excelled at. What a brilliant thing. Innocent intrigue oozed from us.

Now, all these years later, I am confident that fully 99 percent of the people who sat in that room are tenured professors with full job security like I was. And I am also confident that, with their comfortable lives, most of them are golfing several days a week. They are no longer on the edge, no longer eager and questing for new experiences. They've lost the hunger I saw in that orientation room.

There is nothing bad about their lives. They have good lives, I'm sure, and spend their energies on things and people they love. Still, there's something disappointing about losing the bright-eyed wonder, worse yet when that pulls them from the frontiers of their chosen fields... where the fun and breakthroughs lie in wait for the willing.

I might have gone down the same road as my colleagues if I hadn't had a powerful, clarifying moment of innocent intrigue. I discovered my hand, metaphorically speaking, and in an instant, my perspective changed profoundly. I've honed that ability to discover, to see things in front of me in new ways. It has guided me on paths to the accomplishments I've achieved so far in my profession and my personal life.

A PALE BLUE DOT

The profound perspective shift that embedded innocent intrigue into my life was seeing a photo of Earth from space—the iconic image known as the Pale Blue Dot, which was created from data transmitted from the Voyager space probe. It was the result of a campaign led by Dr. Carl Sagan, a brilliant man who lived immersed in innocent intrigue, to turn the probe around so it could view the planet from which it came.

When I first saw that image of a small dot of light in the middle of a belt of thousands, if not millions, of other dots, I was deeply moved. Here we are, living on the surface of a world so immense that from our view-points, it seems flat rather than round, and suddenly we see the planet from millions of miles away—a speck in a massive universe. Two very different perspectives of the same thing. Over time, I've returned to that experience, and it has taught me to make profound perspective shifts in anything I work on. It's a skill I use now for everything I do.

The natural outcome of a profound shift in perspective, innocent intrigue has fostered enormous benefits, and not just for me. Though it may not be identified as such, the history of medicine has many examples of this phenomenon. Antony van Leeuwenhoek is one. A fabric merchant by trade and with none of the training of the established scientists of his age. Leeuwenhoek said, in a letter dated June 12, 1716, he had "a craving after knowledge, which I notice resides in me more than in most other men." That's a pretty good way of describing innocent intrigue, as that craving for knowledge led him to learn to grind lenses, then to build microscopes (inspired by the work of Robert Hooke), and then to train those scopes on a wide variety of subjects. He was the first to see bacteria, protozoa, and blood cells, and for years he shared his observations with the distinguished Royal Society of London. The rest, so to speak, is history.

Alexander Fleming is the name most often connected with the discovery of penicillin, but his was not the innocent intrigue that fully established the value of the antibiotic. A laboratory full of intrigued innocents, including Dr. Howard Florey, Dr. Ernst Chain, Dr. Norman Heatley, and even lab assistant Mary Hunt, acted on Fleming's findings. Through a continuous quest for knowledge and solutions, they not only purified penicillin and proved its efficacy, but found a strain of mold that could produce the vast amounts of the antibiotic needed to address patients' needs. Again, the rest is history.

An example from my own experience was a collaboration with Dr. Robert Dillman, a gifted oncologist who has done years of brilliant research on cancer cures. He was pursuing an idea for "educating" the immune system to kill cancer cells. The methodology involved breaking down tumors, removed during the normal course of cancer care, into single cells. He grew more cells from those he had acquired from the excised tumor and then tried to feed them to the patient's immune cells, in an attempt to get the immune system to recognize and destroy the cancer. His success rate was not high, not even 20 percent, even after 15 years of research. Put simply, the methods he was using to grow the tumor cells were less than ideal, stressing them out so much that the vast majority mutated or died off. At the point in time when I encountered him, the program was being shut down.

When I saw Dr. Dillman's work, innocent intrigue kicked in. I was able to look at the results from an entirely new perspective—with a stem cell biologist's eye. Scrutinizing the methodology used to acquire the cancer cells and grow them, I had one of my "hand-discovering" moments: the only cells, I thought, that could survive such a stressful, inadequate environment were stem cells. A few months of investigation proved I was right. Dr. Dillman had inadvertently purified cancer stem cells.

My realization re-energized the program and led to an entirely new way to make the treatment. Our collaboration—two people with two different perspectives—produced one of the most effective treatments for a wide range of cancer types, using cancer stem cells. Melanoma, glioblastoma brain cancer, ovarian cancer, liver cancer—and this was just the beginning. A new perspective. Innocent intrigue takes the win.

Innocent intrigue is the secret sauce that precedes and catalyzes that other I-word:

innovation. The need for innovation has been catalogued and described in book after book and article after article over the years. Still, it is an unfulfilled quest for many companies and entrepreneurs. Without innocent intrigue—without that eyes-open, ears-open, all-senses-lit-up experience—innovation will remain elusive.

THE AUTHOR OF HUMAN ADAPTATION

There's a biological irony here. The feature of the brain that has made us most adaptable is also the one that shields us from innocent intrigue. The frontal lobe, behind the forehead and between the temples, is the most recently evolved part of the brain. Few animals have a frontal lobe that is well developed. Only evolutionarily advanced species have a well-developed frontal lobe, and humans are one of those species.

The frontal lobe is responsible for forward-thinking and higher-level cognitive functions, including social behavior. It's the part of the brain that keeps us from kissing someone right away when we find them attractive. Other parts of the brain may say, "Whoa, I'm attracted to that person!" while the frontal lobe says, "Calm down, think twice before making physical contact."

The frontal lobe's role in forward-thinking and higher-level cognitive functions allows a person to think ahead and see things from multiple angles. For example, when engaging in negotiations, rather than jumping to the terms of the deal right away, the frontal lobe might say something like, "Play a game here. Drag this out a little so you can explore and learn before you cut to the chase." It keeps you calm, makes you stop and think before acting.

The frontal lobe is the author of human adaptation. It allows you to figure out how to align with your environment and, once that is accomplished, it keeps you settled. For a young professional, it could be operating in their self-talk, "Okay. I've got a house now. I've got a good income, I can pay my mortgage, so I won't lose the house. Okay, settle." Or perhaps, "I've got a life partner who supports me, I'm supporting them, and I have a lot of joy in my days. Okay, settle." The frontal lobe offers the cognitive ability to adapt to challenges as well as settle and stay calm once the adaptations have worked.

The biological irony here is that while the frontal lobe plays a big part in adapting when challenges need to be met, it also encourages inertia once the challenges have resolved satisfactorily. It inhibits the "fight or flight" reflexes of the "lizard brain"—the brainstem, the most ancient part of the brain from an evolutionary standpoint. It got its slang name because it is almost precisely the same structure found in reptiles. At its most basic, it prompts us to feed, mate, or fight, depending on the circumstances. "I'm hungry. I'm going to eat." "Ooh, I like that being over there! I'm going to go mate with it." "This is angering me. I'm going to hit it." The frontal lobe inhibits the brainstem and prevents the immediate reflexive action, allowing you time to consider and calm down.

Your frontal lobe calms and inhibits your impulses. It also helps you look forward, take on challenges, and adapt. Many of us reach a point in our lives when unexpected problems are few, and our frontal lobe reassures our brainstem: "Okay, you can calm down. You don't have to take on challenges anymore. You don't have to fight. You don't have to worry about where to find food."

This evolutionary advantage that helps us adapt so well works against innocent intrigue once we've added years of experience to our lives. Unless we consciously put ourselves into situations that will foster the sparks and the excitement, we won't be able to access the energy and willpower through which amazing things are created.

THE QUIET DEATH OF OVER-SPECIALIZATION

Another aspect of life today that can sap our appetite for innocent intrigue is the need to specialize. The world continues to get more and more complex, and this necessitates the breeding of specialists—successful people who are finely focused. Anybody with a career understands he or she must specialize, and the price of specialization is diversity. The more we specialize in our chosen field, or in our lives in general, the less diverse we become.

Diversity keeps us young and innocent in our thinking, marveling at discoveries, and welcoming shifts in perspective as we learn more about more things. Watch a kid, say, putting toothpaste on his toothbrush for the first time: he's amazed at all that stuff in the tube, the way it squirts out, and how he can squeeze it hard. That's just extraordinary. (I watched my son do this when he was younger, and the image will never leave me!)

Routine and specialization rob us of innocent intrigue and the excitement that comes from experiencing diverse things. It's important to counteract that loss consciously as we get older.

How can an experience-laden adult with a calm and settled brainstem cultivate innocent intrigue? People who are hungry to build and create breakthroughs throughout their lives must avoid the predisposition humans have toward inertia as they age. They must find ways to develop the state of mind from which innocent intrigue can spring.

Here are ways I've found to create that state for myself.

Consciously shift perspective. When I first come across a situation, I have taught myself to change perspective. A way to start doing this is to seek other people's perspectives on the situation. This, however, is not the place to stop. Looking externally is the first step, but the more powerful perspective shifts come from within. Cultivate the ability to shift perspective on your own. Challenge your viewpoint, push yourself to shift to other viewpoints, see the situation from different angles. The challenge or problem will become crystal clear, and unique options for a solution will become easier to see.

Actively seek new perspectives. I challenge myself all the time by seeking new perspectives—any kind of new perspective. I go to educational events or engage in experiences about topics or activities about which I have no knowledge. I have had the good fortune to meet with His Holiness the Dalai Lama, and the experience offered new perspectives in abundance about the heart and mind within us. I've also met a leader of the United Nations and was awestruck by our conversation about the globe we live on. These experiences, as well as the more common activities I pursue as part of my everyday life, give me the fuel I need to shift my perspective, see things from new angles.

There are opportunities to seek new perspectives right where you live—a lecture at your local university, a program at your church, an adult education course, a store that offers seminars around their commercial focus. There are business organizations that provide their members with opportunities for new perspectives as part of their mission. Through such organizations, I've talked to an astronaut about his experiences, listened to presidents and ministers from other countries, and had conversations with a wide range of people I wouldn't encounter in my daily life.

Travel is an excellent way to gain new perspectives. Putting yourself in an unfamiliar culture and engaging with the people you meet...now, *that* is a great way to foster innocent intrigue. That newness can be exciting and exhilarating, and it can open doors to different ways of viewing life and work.

I love going to Indonesia to get a travel-oriented perspective shift. I believe Indonesians are the happiest people in the world. Their standard of living is not at all high, and yet they are extraordinarily excited. It's not unusual for me to come back from a visit there and then watch an executive of mine getting serious and stressed over some business deal that will make him a quarter of a million dollars a year. I'm thinking, "Lighten up. Why are you making yourself so heavy? I just met a guy in Indonesia who makes $5 a month, and he is outrageously happy. You don't need to let your environment destroy your happiness." Travel gives me a new perspective I can bring into my own life.

Hire people who can access innocent intrigue themselves. I like to hire inexperience. For example, I never employ experienced cell culture technicians because I don't want them to bring their technical baggage, their scientific preconceptions, to their work. I want my technicians to come to the work with an attitude that says, "Wow, I'm playing with human cells, and I *need* to make them live in this dish!" That's innocent intrigue, and it helps them think laterally, to invent without even knowing they're doing it. The energy of setting out to conquer an exciting new challenge is limitless. No fatigue, and lots of fresh perspective. People who can access innocent intrigue are "the sky's the limit" type of people. They are more questioning, more willing to shift perspective, and, as a result, are much more creative than their unintrigued counterparts. So, include inexperienced people on your staff and take advantage of the energy and willpower created by their innocent intrigue.

Stay surrounded by young and naïve people. I've already pointed to youth as a fountain of innocent intrigue. Teenagers and early twenty-somethings have an "eyes open for the first time" quality, a "Wow! I've never seen this before!" approach that keeps me from settling into inertia.

I make a point of surrounding myself with young people. I usually have about a dozen interns in my company throughout any one year, some for the whole year, and some just for the summer. For a company of forty people, that's a lot of interns. Their energy and inquisitiveness are catalysts for us, and the benefits are a two-way street. I always make sure a few interns are very young, around 12 years old. They get to hang out with a bunch of entrepreneurial scientists and see that science can be cool, building a company can be cool. My goal is to shatter any preconceptions that scientists are inaccessibly geeky, that businesspeople are somber and reserved. Young kids who are silly bring out our silliness, and they get to see we can be silly while we are doing amazing things. I hope the experience inspires them, that for the rest of their lives they can say, "Hey, I'm just like them. I'm silly, and I can do great stuff too."

The benefit of all these young interns for the company and me is being marinated in innocent intrigue. I maintain an open-door policy so, at any time, one of these interns can come into my office, saying something like, "These cells we're playing with can make a human, and I just played with them in a dish! That is amazing!" I've played with stem cells for three decades now, and it can be easy to forget the wonder of it. It is invigorating to see their perspective. It makes me smile, makes me remember that feeling of innocent intrigue. And it's pretty certain that when I'm playing with those cells the next time, I'll be looking at them with a younger, newer perspective. And the time will fly by—it's been so much fun playing in the lab because I always did so from a fresh and innocent perspective. It becomes play, not work.

It's also important to interact regularly with naïve people. I use that term in its best sense. As Frank Sinatra so eloquently taught us, fairy tales can come true if you are "young at heart." That's what I mean by naïve. It's someone who, at whatever age, can access innocent intrigue, even if it's been a while. I will host fifty-something power brokers, buttoned-up captains of industry, in the lab. I'll show them something fresh, something they've never seen before, and I get to watch their faces and hear their reactions: "Oh, my God! What you do here is incredible. What you've built is so wild. A treatment for cancer! How bold!" Their reactions as they tap into their own innocent intrigue reaffirm my work, re-inspire me, and re-spark my own energy and excitement.

Pursue scary activities. Put another way, go thrill-seeking. I do this all the time: I rock climb, I fly helicopters, I ride motorbikes, I race. I only recently learned to swim, but I throw myself into water sports because it scares the heck out of me. I just started downhill mountain biking, which is hard-core stuff. I've never done it before, and it really scares me. The first time I really fell for my wife was when I watched her jump into an out-of-bounds double black diamond ski run, on a board she labelled "Nik-Monster."

Why does this work? Think back to the frontal lobe I talked about earlier. Losing the inhibition of the frontal lobe, that "well, wait, let's think about this" message, will bring on all kinds of new perspectives. One way to deemphasize the frontal lobe is to trigger the brainstem with something it thinks could be life-threatening. Throw yourself into frightful situations because your frontal lobe is not working when you're afraid you're going to die. The only thing that's working is your brainstem. The rest of your brain shuts off and your brainstem rules your body. It elevates your heart rate. It quickens your breathing. It pumps more blood to all of your muscles, dilates your vessels, and gets you ready to fight or flee.

Losing that "adapt, adapt, adapt, settle" frontal lobe inhibition resets you. You rock climb, for example, and it scares you because you're only thinking about surviving. When you get to the top, you realize your achievement, and it's like you've been meditating for days. Your mind is clear of distractions. Your intrigue pops up and you think, "Wow!" It's innocent. It's new. It's a fresh perspective. If in those moments you start thinking about work, you're brilliant and you're tireless. Your viewpoint is leaping all over the place. You're going deep, and great things can happen.

Share the awe. Take those young and naïve people you've surrounded yourself with and share one of those experiences that used to scare you and that you've now mastered. This is another of those two-way benefit tools for fostering innocent intrigue. They get to experience something scary, thrilling, exhilarating—possibly something they would never have done otherwise—and you get to relive the amazement, joy, and innocent intrigue of that first-time experience.

For example, I have a helicopter, and I take people flying all the time. They get to experience being aloft in a helicopter, looking at the world in a new way. I watch their expressions as I fly them around and listen to them exclaim as we pass over a whale, as we zoom the beach at 200 feet and people wave, as we crest a mountain peak and the land suddenly drops away below us. I hear them yelp with joy, and it makes my flying experience more fun. It recreates that sense of perspective, of new wonder and limitless energy for me. They have a fantastic experience, and I get to access innocent intrigue through them.

Embrace fun in daily life. Being in environments and situations that are fun creates a limitless source of energy, because all the things that are dragging life down are forgotten. Laughter is so helpful to tap into innocence. When we laugh, we are kids again. Our brains are awake, alive, uninhibited. When we get to that space of innocence, the intrigue isn't far behind.

If I have friends who are feeling stalled and feeling heavy, I do laugh-inducing things for them. I send funny videos or tickets to a live comedy act. I get silly; I tell them jokes, and it pulls them out of their stall. I may not be the funniest guy, but I can be as silly as the best of them.

Another way to embrace fun is to play with a child in their environment—on the floor, in a sandbox, at a playground. Immerse yourself in the play, and you'll end up smiling. It's really, really fun.

Having fun nurtures a sense of tireless energy in you that leads to innocent intrigue. It pops you out of any heavy and significant perspectives you may have and gives you a fresh outlook. Make sure you embrace fun. If you don't, the odds are you'll become a boring and uncreative businessperson.

Look for innocent intrigue yourself. Know what tools you can use to foster innocent intrigue and keep a tool list handy, so you can remember to use them. I've offered some ways I generate the energy and willpower from which I create and innovate. I invite you to come up with your own strategies. Make your mantra "Bring on the innocent intrigue!" Look for tools and write them down. Put them on sticky notes and tack them up around your home and office—your desk, your bathroom mirror, your refrigerator, wherever they will be seen.

I have lists I refer to from time to time to remind myself. "Oh, yeah. I forgot about embracing fun." Or "It's been a long time since I have jumped off a cliff; I really need to do that again." And I'll take action to get those things into my life. Sometimes I'll meet somebody who has done something wild, something I haven't thought of. And that experience gets added to my list.

Innocent intrigue produces free energy, free willpower to fuel breakthroughs. There's no limit to it. It's like love: there's no end. You fall in love with somebody and you think, "Wow! I'm in love. This perfection needs nothing more." Then you have a child, and it's, "Wow! There is even more love in my life." It's the same with energy and willpower.

Innocent intrigue fosters this energy and willpower, and that is the path to creativity and breakthrough. But the only way to achieve innocent intrigue once you've become an adult mired in preconceptions is to act on purpose to access it. Increase the odds of landing in innocent intrigue. Keep drinking from that well, and you will create an unlimited source of pure, unadulterated energy and willpower with no preconceptions.

And then watch out world—amazing things are about to happen.

2—TRY (JUST SHUT UP AND DO IT!)

In his poem titled, "My Heart Leaps Up," William Wordsworth wrote, "The Child is the Father of the Man." This line, as well as the context of the poem, has left an impression on me over the years. In it, Wordsworth rejoiced that the wonder and optimism he felt as a child had remained alive and well within the iterations of his adult life.

This sentiment, celebrating the joy of childhood which has survived within the adult, resonates deeply with me. In many ways, my own childhood experience growing up in rural Canada has defined and laid the foundations for the way I live my life.

LEANING INTO THE HARD STUFF

The south-central section of the province of Ontario, where I grew up, has the kind of weather one would expect for a northern latitude of around 44 degrees: mild, balmy summers followed by nine months of frost-bitten fingers and deep snow. As part of a hard-working family with exceedingly little money, my five siblings and I participated in a vast array of chores on the farm.

From the age of seven until my eleventh birthday, I had a very specific job in the family: milking the dairy cows. Rain, sleet, or snow, it was my job to get those cows corralled into the pen and milked. To the uninitiated (i.e., those who did not grow up on a farm) this might not sound like such a Herculean task.

However, to a scrawny seven-year-old boy, all elbows and knees, clearing a path through that bovine mess proved to be something of a challenge. Given my small stature, the cows were completely unintimidated by me, and I had to elbow them in the belly, left and right, just to make any sort of headway. Meanwhile, I was to track down the dairy cow that needed milking, begrudgingly pull it into the milking pen, and then get the half-ton beast to stand still while I messed with its udders.

After the day's milking was done, the real work began: shoveling s—t. Cleaning out the pens meant removing the endless piles of manure—work that went on until I could hardly lift my arms. Season after season, my brothers and I elbowed cows, shoveled s—t, picked potatoes until our hands were raw, and handled chickens until we temporarily lost our sense of smell. It was backbreaking work, and it was our way of life.

Just shut up and do it. That was a large part of my childhood experience. No room for drama, just a task that needed to get done. It's a work ethic, an approach to life I internalized as a young boy and one that has continued to serve me to this day.

But let me be clear—I had a great childhood! Any pity felt for the poor farm boy with too many chores would be misplaced. I was extraordinarily happy as a child and herein lies the lesson: children don't infuse a particular task with preconceptions about whether it's good or bad, they just do it. More often than not, they even end up turning that activity into something pleasurable and entertaining.

Growing up on a farm in a big family is terrific. I'm pretty sure I decided that I wanted to raise my own children on a farm before

I even knew where kids came from. There's a lot of love and a lot of freedom, and the playground is immense. Getting lost in cornfields, trying to ride cattle—which, by the way, is impossible—and lots of other fun things to do made my life immensely fulfilling amidst all the hard work. Like Wordsworth, the laughter and love from those days are with me still, and this aids me every day in my work.

I chose a vocation that can be heart-wrenching, and without tapping that sense of childhood joy, it would be difficult to see past the hardship of the patients I interact with, and the enormity of the tasks my team takes on. I feel fortunate to have been trained to do the work that has to be done, and to do it with a light heart. I've been lucky to find the joy, the fun, and the satisfaction in getting things done without all the stress and heartburn so many people bring to their work.

JUST DO IT

Master Yoda once said, "Do...or do not. There is no try." While that might be good advice if you're going to levitate a TIE fighter jet out of a swamp, it becomes noticeably less helpful in the rest of our human endeavors. Where Yoda and I diverge in thought, is that the word *try* is not a passive verb in my world. It is part of doing–the actions that ultimately lead to success and accomplishment. In my case, there is no *doing* if there is no *trying* first.

My father was a living example of putting the word *try* to use as an action verb. He first demonstrated this to me as a farmer, then as a pharmaceutical company executive, and later as a company CEO. In my late teen years, our life took a dramatic turn. Put simply, we went from being poor farmers to the family of a Big Pharma executive. I didn't know it at the time, but my father had spent the early years of his career as a lab technician, and he had become familiar with Big Pharma by way of the salespeople who came through the lab. Because he had never entirely severed his connection to the pharmaceutical industry, he was able to earn his way to an executive position at a large company when I turned 18. He stayed with that firm through several structural iterations before striking out on his own at a series of small companies. Some were successful, others were less so, but all were a result of my father trying, trying, trying— and occasionally succeeding.

When it came to my education, the rule of the farm seemed to work there too—*shut up and do it*. I did, and I've never stopped. My childhood education was what you might expect in a poor, rural community: one teacher to every three hundred or so students. Luckily, I had a decent head on my shoulders and a wealth of curiosity, and I made the best of my situation. When I set my sights on a university education, though, I understood the financial challenge and realized my parents would be unable to help me.

My childhood education hardly prepared me for the true cost of a degree; after doing some research, I was blown away by the figures. How the hell was I going to fund a four-year degree?! After probing the rabbit hole of higher education for some time, I found only one logical conclusion: scholarships. I needed government money—and to earn it, I would need grades that would merit regional, possibly even national, attention. I made up my mind right then and there to become the top student in my class. I would apply myself. I would try and try and try. If I didn't, I'd never be able to afford the education I wanted.

Since there was no foreseeable way around it, I became an excellent student. I furiously applied myself and remained at the top of my class all the way through my educational career, even into my PhD. I never stopped trying, and I never allowed myself to get comfortable. Scholarship after scholarship, I did what I had to do to guarantee I would get the funding that would allow me to progress further.

The year I received my PhD, I was honored by an award for the most outstanding PhD thesis in the country. That award helped secure my admission, as well as a full-tuition scholarship, to the University of Cambridge for my postdoctoral studies. I was both overjoyed and relieved—but, truthfully, my academic performance was born from something surprisingly simple: necessity. When the going got tough, I just had to work that much harder. There was no other option.

Ask any veteran of a PhD program and you'll hear battle stories about wars won and lost at the culmination of their final dissertation. It's a tough life, one with very little room for vanity or ego. There's a lot of trying, and there's even more shutting up and doing it, because that's what's needed to reach the goal.

I'll be the first to admit I was a bit of an overachiever. Many people get scholarships for their postdoctoral fellowships without breaking their backs to earn top awards as well. For me, though, there was simply no other alternative. The issue was black and white; if I didn't get the funding I needed, my research was over. Determination molded my approach, the same determination that propelled me through a herd of cows when I was young. In my mind, getting anything less than top grades and honors meant a chance I wouldn't receive a scholarship. So, I consistently put myself in the top zone where funding was guaranteed.

ON THE BACKS OF GIANTS: MY MENTORS

Over the course of my career, I've been inspired by many people who embody the quiet philosophy of simply shutting up and getting it done. Of those silent champions of hard work and effort, two people in particular stand out to me: Rick Hansen and Christopher Reeve. Both men sustained spinal cord injuries that could have easily led them to victimhood and feelings of self-pity. Instead, they made amazing contributions toward spinal cord research and actively fought to improve the quality of life for people living with disabilities.

Rick Hansen is a Canadian treasure. As a young man, he was a fantastic athlete, winning awards across multiple sports. A spinal cord injury, the result of a truck accident, left him paralyzed at age 15. From the beginning, he did what he had to do to adapt to his new situation and get on with his life—he just shut up and did it.

Rick graduated high school and went on to university, where he earned his degree in physical education—the first student with a physical disability to do so. Despite being wheelchair bound, Rick was a natural athlete and continued to pursue his passions. Over a 5-year period, he would win 19 wheelchair marathons and 15 medals—6 at the Paralympic Games and 9 at the Pan Am Games. He gained worldwide recognition with his Man in Motion World Tour—a two-year journey around the globe in a wheelchair—which raised $26 million for research related to spinal cord injuries.

To this day, Rick continues to be a man in motion: he is the founder and leader of the Rick Hansen Institute, a non-profit organization that supports spinal cord injury research, accessibility, and inclusive educational policies. He has repeatedly been a source of inspiration for me throughout my career.

Christopher Reeve is another such man.

Although I first knew him on screen as The Man of Steel, it was Christopher Reeve *the individual* who would ultimately leave a bigger impression on my life. In 1995, the highly successful actor sustained a spinal cord injury from a horseback-riding accident and emerged a quadriplegic. Like Rick, Christopher did not surrender to his situation. He became an activist, working on behalf of people with disabilities related to spinal cord injury, and proved himself to be a true man of steel.

Christopher was an early proponent of stem cell research for spinal cord injury. In fact, with Joan Irvine, he co-founded the Reeve-Irvine Research Center, which continues to be a global leader in spinal cord injury research. I had the great honor of being the first professor to be hired there by a newly appointed director, and I got to know Christopher pretty well during my tenure at that institute. His unwavering optimism and down-to-earth attitude were infectious. It was impossible not to be inspired by him. Of course, he also was a grandmaster at the art of shutting up and getting things done, typically in the most amazing ways.

Often, Christopher could leverage his celebrity status to bring media attention to important issues, thereby making positive, quantifiable changes. In one instance, he appeared on *The Tonight Show* (then hosted by Jay Leno) and remarked on the backward logic of insurance companies that were unwilling to pay for two ventilators beneath a quadriplegic's chair. If one ventilator were to fail, Christopher noted on live television, that person in the wheelchair would die. The next day, on *The Oprah Winfrey Show*, he told the follow-up story: his insurance company contacted him after his appearance on Leno and informed him they would cover his second ventilator. Christopher's response? Not a chance. He told the insurance company, "I'm not going to do it until the insurance policy changes, and everyone gets one."

A week later, the policy change went into effect.

Hunger strikes were another useful tactic in Christopher's "make a difference" arsenal. It was not unusual for me to talk to him and find out he was in the middle of a hunger strike. If I expressed concern, he would invariably tell me, "Oh, it'll be over in two days. Just watch."

One of my favorite memories was born the day he politely summoned me to L.A. I arrived at LAX and found he had booked the entire top floor at one of the airport hotels. I was escorted into his room, and we chatted. He was on a hunger strike and had strategically positioned himself to be accessible to the press. He had invited reporters to fly in from across the world to interview him, and he had a string of meetings lined up that morning into the afternoon. In that particular instance, his focus was on a bill he wanted to be passed into law.

"Chris, you look terrible! Are you OK? What is going on?" I asked, stepping into his room and full of concern.

"Oh, I'm on a hunger strike," he replied nonchalantly. He was full of energy and apologized for it. "I get kind of hyper when I haven't eaten for a day and a half. I'm just bouncing off the walls."

I was totally awestruck and told him as much, "You're a freaking machine!"

He just laughed and said with absolute conviction, "Oh, I'm a freaking idiot. But you watch, man, I'm going to do this." And you know what? He did. He got that bill passed.

A man trapped inside of his skull, with no use of anything but his brain and face.

Rick Hansen and Christopher Reeve shut up and did it. They knew what impact they wanted to make in the world, and they did what they had to do to reach their goals. I couldn't fully appreciate the enormity of their spirit before I began my own career working with people who live with spinal cord injuries. After seeing their lives up close, however, I can tell you they are some of the strongest and most resilient people you could ever encounter. Whenever I catch myself feeling pity or thinking of my patients as victims of circumstance, I remind myself of the time I spent with Christopher. I remember the smile that never left his face as he stubbornly went about changing the world.

Just shut up and do it.

This can be a real challenge for a leader. Sometimes the going gets rough for me, and my natural instinct is to turn away from the hardship and seek comfort. To do so would not only be counterproductive, but it would also go against my life's work. This is when I remind myself to *shut up*. Rather than forcing myself through sheer will or trying to strangle my monkey mind into submission, I choose to practice mental clearing.

MASTERING THE MENTAL RESET

There are myriad ways to center oneself, and different methods work for different people. Personally, I draw from my experience in martial arts. Whether I'm fighting one or more senior belts in tae kwon do or confronting a table of board members, I visualize a calm pond of water in my head that will not be disturbed by distraction or digression. Qigong (pronounced *chee-gong*) offers similar concentration techniques. Qigong is a subdiscipline within tai chi focused on moving energy within your body for the purpose of healing yourself or others. In qigong, to quiet your own mental chatter, you concentrate on a point between your eyes, a centimeter up, and about a centimeter inside your head.

Try it.

Close your eyes and direct all of your attention to the area above the bridge of your nose, right in the middle of your forehead, and about a centimeter deep inside your head. You'll find it's easy to shift your attention to this area of your body, and it can be a powerful tool for slicing through the Gordian knot of distraction. When I need to quiet my mind and center myself during a crisis, I find a quiet place to sit and mentally shift my focus to that point. When I can take three long breaths without a single thought entering my mind, I know my brain is clear. Truth be told, I have often done this while in the hot seat, in front of a dozen stressed-out and babbling executives. Although I may miss a few sentences, I will own the rest of the meeting.

Similarly, some people prefer to use positive affirmations before going out and getting the thing done. It's like chanting a mantra. Look at yourself in the mirror and affirm, "I've got this. I'm going to listen and think clearly." Or "I own this issue." Focusing on an affirmation can give some people the springboard they need to clear their minds and follow through on the *doing it* part.

Regardless of how you find peace and clarity in your own life, one thing remains absolutely certain: the strongest will in the world, left unmanifest, does absolutely no good for anyone. Manifest will—ideas and goals made real in the world—is a balm for the soul and an engine for progress. There will be obstacles; and on some days, there will appear to be unclimbable mountains. To realize those ideals and reach the level of personal growth that represents an authentic expression of yourself, you have to *try*—so shut up and start doing it.

3—PEOPLE ARE EVERYTHING

In the last chapter, I encouraged you to get rid of your noise, cut through everyone else's noise, and just shut up and do it. But you'd best be sure you are happy with the answer to one simple question before you start down that road: *Why* did you set out to do that thing in the first place? Having taught PhDs and MDs for much of my career, I can tell you with certainty that the most common answer—"to make other people proud of me"—is a catastrophically terrible reason to be doing what you are doing.

WHAT IS YOUR "WHY?"

The vast majority of high-functioning people have never really considered what their primary drivers are, why they do what they do. For me, the answer has always been obvious: people. People are quite literally everything; they are my motivation, my inspiration, and the drive to continue on in the face of ridiculous adversity. I've been honored to have met many incredible people in my life, certainly more than I can hope to fit into one chapter. That said, I'd like to share a few of the more memorable encounters and why they underscore that people are everything.

Though I've introduced them already, any discussion about heroic figures in my life would be incomplete if I didn't briefly mention Rick Hansen and Christopher Reeve once more. Although Christopher drove my desire to find an effective treatment for spinal cord injury later in my professional career, it was Rick Hansen who first showed me the path.

I had seen Rick on television many times, and I met him in Vancouver after his return from the Man in Motion Tour. Even back then, before I really knew him as a person, he blew me away. As a young boy, I was amazed by the feats of this incredible athlete, this Olympian. He might have been wheelchair bound, but my young eyes barely noticed—the man looked as if he had been forged from steel! Even more than his physical appearance, Rick impressed me with his incredible displays of will. We're talking about a man who had taken it upon himself to traverse Canada *in a wheelchair*! When that didn't prove challenging enough, he decided to keep going...and pushed that wheelchair around the world.

I wasn't aware of this at the time, but in Rick Hansen I was watching the personification of my martial arts training—an indomitable spirit and relentless perseverance. Rick was the real deal. It wasn't just his musculature: he was relaxed, articulate, energetic, and fun to be around. Deciding to push a wheelchair around the world had never been about ticker-tape parades or marching bands; he just wasn't that kind of guy. One thing I didn't appreciate until years later was that I had never, not once, thought of him as a paraplegic. In fact, that always seemed like an unimportant afterthought. When I looked at him, I saw only a champion of spirit and a world-class athlete.

In the course of my research, I have learned a great deal about the daily lives of the people for whom I was developing a treatment. When I see someone in a wheelchair, the scientist in me immediately assesses their *deficit*. At a glance, I can quickly assess the state of their arm and leg muscle function. My training also allows me to see signs that suggest sensory loss, pain, and spasticity, as well as bladder, bowel, and sexual dysfunction. I then mentally translate those deficits into that person's lifestyle. I consider how demeaning it can be to have a nurse assist during a bowel movement. I think about how personally assaulting it is to smell like urine because your urine bag is leaking and needs to be managed. I notice all the granular challenges of the day—that bumpy sidewalk, the daunting flight of stairs, that small door to a stall in the public bathroom, that narrow aisle in the airplane, or even the temperature fluctuation that can cause a sudden and crippling short circuit in their injured nervous system.

Rick never showed any signs of discomfort with the challenges life threw at him. Instead, he tackled the world, changed policy, and inspired over a billion people. To his credit, the whole world saw what I saw: a hero, a winner, and a guy who was on top of the world.

Nothing in my life is as heavy as what Rick and other patients with spinal cord injuries face every single day. Yet, for all their added difficulties, I consistently see overwhelming optimism and positivity in that community. Take, for example, the elderly gentleman I was helping at an event for people in wheelchairs. This man had no motor function below his chin, except for some very minor control in one of his thumbs. With a Cheshire cat-like grin, the man smiled at me one day and told me how extremely lucky he was. I looked at him incredulously and asked, "Why?" Not understanding my confusion, he gently explained, "I can wiggle my thumb to make calls, to drive, to have some independence. That poor man over there can't. I am forever reminded of how lucky I am."

These are the people who have energized me, who have reinforced my determination to develop medical treatments, and who have spurred me on to manifest my own indomitable spirit better.

Christopher Reeve was another such man.

During my postdoctoral fellowship at the University of Cambridge in England, my research had received some acclaim because I had made a few seminal discoveries regarding the control of regeneration within the brain and spinal cord. This, I believe, is how I came to the attention of Christopher Reeve and Joan Irvine. They contacted me, and after a dance of interviews and negotiations, I became a professor working at what was to become a leading center for spinal cord injury research, the Reeve-Irvine Research Center at the University of California at Irvine.

It was during my very first week as a brand-new professor, while sitting in my brand-new office, that a phone call rocked my world.

A breathy voice, decreasing in volume as the respirator recharged, whispered into my ear, "Dr. Keirstead, this is Christopher Reeve. I just wanted to congratulate you on your new job."

I hit the mute button. If anyone had peered into my office at that moment, they would have noticed a dancing young scientist flying about the room and screaming, "Woohoo! Superman! Superman just called *me*!" Christopher Reeve was a childhood hero of mine, and it was a phenomenal thing to hear from him personally.

That initial phone call was the catalyst for an amazing collaboration and a great friendship with Christopher.

When he passed away in 2004, I was grief stricken. I was also angry at myself—angry I hadn't worked fast enough to find the treatment that would have given him mobility. His passing strengthened my determination to succeed, and my achievements in the treatment of spinal cord injury were motivated to a great degree by him. That guy, with his smile and his energy and his dedication, made an enormous difference in my life and work, and I revere him to this day.

There's another person who fundamentally influenced me in my young years: Carl Sagan. In my eyes, Carl Sagan was not just a man—he was *the man*. A multidisciplinary scientist, Sagan was an astronomer, a cosmologist, an astrophysicist, an astrobiologist... and more.

It was Carl Sagan who put together the universal messages that were sent in the Pioneer and Voyager probes. And of particular significance for me, it was Sagan who had NASA turn the Voyager probe around to capture the image of the Pale Blue Dot—the one that zapped me with innocent intrigue.

Science is incredibly pervasive, touching on every single aspect of our society. Through his television programs and books, Sagan accomplished something incredible; he brought science to a lay audience in an accessible way, helping them see the world and the universe in new ways (hello again, innocent intrigue!). I loved seeing a serious scientist, someone with ironclad credentials, become a popular media personality who influenced the world.

Though I never had the opportunity to attend one of Sagan's lectures or meet him in person, he nevertheless left an indelible impression on me. In Sagan, I came to realize a scientist could be so much more than an academic, and that science was a way of life rather than something that you left at the lab when you went home for the evening.

While attending the University of Cambridge some years later, I had the great pleasure of meeting the theoretical physicist and cosmologist Stephen Hawking. Stephen had been diagnosed with amyotrophic lateral sclerosis (ALS), a motor neuron disease, in his early twenties and, as an adult, was wheelchair bound and only able to communicate through a speech synthesizer. Here was yet another brilliant mind trapped inside of a cranium, moving with an intellectual athleticism to conquer the world. He liked being wheeled by the River Cam, or just parked to watch the birds and the river and the punts go by as he thought. Whenever I saw him, I'd walk over and have a chat. Like Sagan, Hawking inspired young people to enter hard science as a field of study, and his work took astrophysics out of the ivory towers of academia and into the mainstream.

In my emulation of Sagan and Hawking, I've also come to appreciate an often-overlooked facet of their careers: they both gravitated towards youth and teaching. After experiencing this for myself, I completely understand why—young people are a closed loop of inspiration. They motivate and recharge me, and in turn, I challenge them to think about the possibilities and wonders of the universe.

During my experiences as a professor, I loved mentoring students and watching their eyes light up when a concept or theory finally clicked. To this day, I continue to hire young people for the internships offered at my company and frequently speak at high schools around the nation.

When I walk into a high-school classroom, I see students with dwarfed attention spans who are hyperactive, overstimulated, and under-challenged intellectually. They are not settled, neither internally nor externally.

When I talk about science, they pay attention mainly out of politeness. Yet, as we touch on concepts such as will, tenacity, and the boundless possibilities of science, I see their eyes light up. When I look out at that sea of faces and think about the total cumulative power of all those bodies in the room, it blows my mind. I feel the potential emanating from them, untapped and unrealized, and it inspires and motivates me.

EVERYONE HAS SOMETHING TO TEACH YOU

Sometimes inspiration can come from the most unlikely places.

At one point in my professional career, I agreed to an interview with CBS's *60 Minutes*, despite the chagrin of the powers that be at the university where I worked and their counsel against it. Without question, it was a risk—at that time, stem cell research was embroiled in misunderstanding that had created some major controversies. Still, I decided to take a chance, and, in so doing, I met an inspiration.

Ed Bradley was a well-known journalist and news professional, and he would be the one conducting the interview. He and the production team made the trip to my laboratory on the University of California,

Irvine (UCI) campus and turned it into a Hollywood movie set. It was a novel and very exciting experience for me (hello again, innocent intrigue!).

As we settled into our chairs and began the interview, I couldn't help but notice Ed was sitting extremely close to me. In fact, he was so close, one of his knees was positioned between mine. I was mildly uncomfortable and couldn't help but wonder why this guy couldn't give me a little breathing room. It didn't take long to figure it out.

During his commanding and authoritative narrative on camera, when he wanted a surprised reaction to one of his questions, Ed would jiggle his knee, prompting a star-tled look from me that would be caught at just the right time. As I am not accustomed to having a man jiggle his knee between my legs, it worked. Ed was a master at his craft, and that was my first view into one of the many silly secrets behind TV interviews. Yet, it was not this interview trick that had me preoccupied. Just before we began, Ed had told me he was dying of cancer. In fact, this turned out to be his last show.

Though he died a few months later, I've never ceased to be impressed by this man, rocking it right up until the very end. Ed is one of several people who have posthumously motivated my work on a cancer treatment.

Sally is another.

Sally was a cancer patient who had been admitted into one of our rounds of clinical trials for a new cancer treatment my team and I had developed. During these clinical trials, the pool of participants was divided into groups—some groups received the test drug, while others received a placebo. This was all done blind, so neither I nor any of the medical staff performing the trial knew which patients belonged to which group. In the trial Sally participated in, there were three groups, and one out of every three patients was unknowingly taking a placebo.

Sally came to visit the company during the trial, and she was so excited to be participating. We gave her a tour of our facilities, and she bubbled over the whole time she was there, infecting my staff with her energy and blinding optimism. Our work had given her new hope, and she was sure she was going to beat her cancer with our help.

We couldn't have known Sally was in the placebo group.

When she passed away, I was at a low of despair and a height of anger, an echo of my reaction to Christopher Reeve's death. This time, my anger centered on the disease, and it fueled my resolve. I have tried to use her sad passing to motivate myself to ensure something good came from this grief.

Of course, much of my energy and motivation comes from uplifting and inspiring cases as well. Christopher from Australia was one of our first melanoma patients. He flew all the way to the United States just to say, "Thank you, Hans, for keeping this treatment and company going. Your crew treated me a decade ago, just after I had been told I had three months to live. And I'm alive." He's well and in the clear.

Another time, there was a knock on the glass entry door of our laboratory offices. I needed a stretch, so I got up from my desk and went to the door. A woman stood there holding a tiny baby, a newborn. With tears streaming down her cheeks, she told me that, six years ago, she had been told she had three months to live because of her cancer. She had received our treatment, had been cancer free since then, and was now a new mother. She could watch her child grow, and that is a beautiful thing.

Though these stories illustrate a small fraction of the satisfaction science has given me, I can't express in words the gratitude I feel to have played a hand in the recovery of these beautiful people. It is exactly these kinds of stories that remind me why I'm not sipping martinis in retirement, at the wrong time of the day, on a fabulous beach somewhere. Hmm...

Of course, none of this would be possible without a very special group of people. My staff is yet another incessant spring of inspiration, so much so that it has become the foundation of my relationship with them. I am a serial entrepreneur who has built and sold several companies, and about half of the people in my current companies have worked for me in prior iterations; some have been with me for decades. We have supported and inspired each other over several meaningful missions to combat disease, and I'm very humbled by that.

Biotechnology is complicated, with a seemingly incessant stream of challenges that requires mastery of complex research tools. The best researchers are often the ones who simply know the most stuff, having lived longer and read more research articles. But such individuals seem a dim flame to the fiery brilliance of a tight team of scientists who know one another's strengths. How many times have I walked heavy hearted into a meeting, fearing our ignorance or an incomplete data set would destroy our ambitions, only to have one of my staff innocently bounce through their latest discovery, unknowingly saving the day, the program, and the company? How many times have I thought, "We are finished, because that can't be done" only to have one of my staff correct me with a smile, and tell me they just had just done it? More than I can count. You may be good at what you do, but a team is better.

One of my favorite examples is my fellow scientist and right-hand man, Dr. Gabriel Nistor. When I met him, he and his wife, along with their newborn baby, were living in an unfinished basement. They had almost no money. In fact, I later found out Gabriel had even shown up to our interview wearing someone else's suit. His family had won a green card, otherwise known as a Diversity Visa, which took them from their Romanian homeland to the U.S. With the admirable and dogged determination of a man who faced challenges and would not accept failure, and with a light and ambitious heart, Gabriel took any and every job he could find to feed his family.

When he told me about his journey to the U.S., I hired him on the spot. Not because of his skills, but because of his character and tenacity. He and I have since worked together for over twenty years. Together, we have created companies, explored new avenues of science, and had the opportunity to save many lives.

Getting to act as a springboard for the betterment of my employee's lives is one of the unanticipated joys of my position. While I have certainly seen some of my staff ignore or reject the opportunities available, I have also seen the majority run, leap, and bound gracefully into happier, more financially stable lives. It is entirely the athlete, not the springboard, who deserves the credit here. But... it would be a bold-faced lie to say that I don't love to watch them soar.

YOU DON'T HAVE TO LIKE THEM TO LEARN FROM THEM

When I assert that people are everything, I mean the whole 360-degree picture. There are people who inspire and motivate in positive ways... and then there are people who provide, let's say, *counter inspiration*. This group of *teachers* behaves in one or more ways that run counter to the basic tenet of this chapter: they teach you what *not* to do. While this group doesn't enthuse, they certainly motivate, and the lessons learned from them are no less valuable. Well, maybe just a little.

In order to protect their professional reputation and so as not to offend anyone unnecessarily, I've changed the name and personal details of the individual in the following story. The lessons I learned from this experience, however, remain true.

Oscar always seemed to be a victim of his giant pulsating brain and his laser-like career focus. He and I were colleagues at the University of Cambridge, where we were both pursuing our postdoctoral fellowships. We were motivated, ambitious, and surrounded by the brightest and most brilliant professors Cambridge had to offer. Oscar was an early adopter of stem cell research and had a lightning-fast career that landed him a professorship at a young age at a very well-to-do university. From there, he rose to direct one of the earliest (and best) stem cell centers in the world. For almost a year after Oscar left Cambridge, I remained buried in my research, focusing on the science rather than my career.

After leaving Cambridge, I took a professorship at University of California, Irvine. Sometime later, I ran into Oscar at a Society for Neuroscience conference in San Diego. At this point, I had only a budding interest and knowledge of stem cells and their practical applications. When I ran into Oscar that evening, he was smug and smiling to himself.

"Hans! What are you doing here? Coming for the free food and drinks?" he asked in a patronizing tone.

When I told him of my interest to begin working with stem cells, I could see his face twist into a smear of mockery and derision. All these years later, I'll never forget his response: "What, now that you've risen to the top of the spinal cord field, you're just suddenly going to become a stem cell expert?!"

While that is precisely what I did, Oscar's arrogance and perspective was, for me, the antithesis of what a scientist, a discoverer, an explorer, or an entrepreneur should be. By categorically placing me in one specialized box, he revealed his underlying bias— namely that a researcher has no business asking questions or showing an interest in any field outside his or her own quiet niche.

It's worth mentioning that Oscar is still a professor at that same university, using the very same approaches he used when he left Cambridge. Like most research, his work has consumed decades of time and tens of millions of dollars. While many thousands of rats have benefited from his therapy, as of yet, not a single human has.

Intellectual blinders have their purpose. I use them every time I'm consuming research articles or trying to learn a new discipline. However, I can tell you from experience that they will crush your wonder, creativity, and zest for learning if you leave them on too long. More often than not, if you keep your eyes and your ears open, feed your innocent intrigue, and challenge your own judgements, you can pass right by those people with their blinders on—and usually with a lighter heart.

While Oscar taught me a valuable lesson about shifting my perspectives and biases, it was one quirky professor in particular who made it clear to me exactly what I did *not* want to be spending time on.

As a new professor at University of California, Irvine, I founded a stem cell research center and even got to help design the new building that would house it. What I didn't realize at the time was that I would also be responsible for running this new research center. While it was an exciting opportunity, the daily meetings were endless. Because stem cells are so ubiquitous in our bodies, stem cell research applies to nearly every human disease or malady. As the head of a new and leading stem cell center, people from all over the world were knocking on my door—literally! Usually, we could help them in some way—sometimes by giving them the cells to work with or by teaching them how to grow and cultivate the cells in their own labs. All of this was wonderful, except it meant a barrage of daily meetings that took me away from my own important work. Still, I shouldered on because I believed this was expected of me.

On one occasion, a doctor flew all the way from India to meet with me—it was my eighth meeting that day. This doctor politely asked if I could show him how to inject stem cells into a very small nerve. "Certainly," I told him, "We can make extremely thin glass needles for fine injections. So, which nerve and why?" Without a trace of humor in his voice, the doctor replied he would like to inject stem cells into the penile nerve to treat erectile dysfunction. I was stunned. Why would you do that and, more importantly, who would want it?! Haven't you heard of the little blue pill? A few days later, I resigned from my position as the director of the stem cell center. It was time to get back to my research.

In some ways, that extremely thin glass needle was the straw that broke the camel's back. Never again would I allow myself to be pulled away from my optimum zone of execution, the place where I know I have a shot at making a real difference. This interaction reinforced my conviction that my time must be spent on science that is clinically and commercially relevant, that will make people's lives better. Science that will save lives.

WHAT WOULD HIS HOLINESS DO?

Outside the scientific world, I have met and worked with many leaders in business and government all around the world. Several of these people have become exemplary sources of positive inspiration. Individuals who actively seek their own way to *be the change*.

Tom Tait is a member of a CEO group to which I belong and an all-around great guy. He was involved in the city governance of Anaheim, California for many years, culminating in an eight-year stint as the city's mayor.

Tom's mayoral campaign theme was unique and really fantastic. It started when he was walking in town one day and saw a little girl planting a sign on the front lawn of her house that said, "BE NICE." He stopped and talked to the little girl, and from that conversation his campaign platform was born.

Kindness.

Tom was elected and re-elected, putting this theme into action. He built a program called "A Million Random Acts of Kindness," launched a mayoral website that allowed people to post their random acts of kindness, and the city watched that list grow year after year. There were real and important results from Tom's kindness programs. The crime rate went down. Schools reported that incidences of bullying were fewer. Respect for the police force increased. Benefits manifested themselves in all kinds of unforeseen ways.

Lama Tenzin, a senior Tibetan monk who directly serves the Dalai Lama, approached Tom about meeting with His Holiness. Lama Tenzin's job is to find people in the world with whom the Dalai Lama can connect, and then to facilitate an introduction. Through Lama Tenzin, Tom met His

Holiness and has drawn on that source of inspiration to better his own life and the lives of others in countless ways.

Lama Tenzin learned of me and my work through conversations with Tom and another mutual friend, Robert. He reached out to me in the fall of 2019 with an invitation to meet with His Holiness. I accepted with equal parts excitement and humility. What an honor!

When I arrived in Dharamshala, I was greeted by Lama Tenzin and agreed to meet him for breakfast the following morning. When I arrived, I found him with four other very notable figures—three theoretical mathematicians and a quantum physicist. Lama Tenzin prepared us for our meeting, and together we walked toward the palace of His Holiness the Dalai Lama.

What had been scheduled as a 45-minute meeting lasted for four hours.

His Holiness and I really hit it off. Our affinity has many sources, but I think the strongest was a shared delight in innocent intrigue and laughter; it seems in retrospect that we giggled and danced our way through hours of deep conversations on science and consciousness. It was such an enlightening and interesting conversation, and it was the beginning of what I'm sure will be a long-lasting and very fulfilling friendship. He is the type of inspiration I carry with me every day. When I find myself stuck or out of sorts, I pause to think, "What would His Holiness do?"

****

People are everything. Every pivot, every motivation, every inspiration in my life comes from individual relationships with people. Respect them, make yourself open and vulnerable to them, be inspired and excited and fascinated by them, and you will find an infinite spring of motivation.

4—BALANCE IS THE OTHER EVERYTHING

Three months into my professorship at University of California, Irvine and I was ready to throw in the towel. After years of dogged determination and long hours in the University of Cambridge labs, I had finally earned a professorship at a reputable university where I could continue my research—or so I thought. To the rest of the world, my life looked pretty full. I was at a prestigious research center, a professor on track for tenure, surrounded by brilliant colleagues and faculty alike.

I was also bored stiff.

However it might have looked to the casual observer, my equilibrium had been thrown wildly out of balance. Landing a professorship was undoubtedly credentialling, but being a professor was insanely monotonous, allowing for no balance whatsoever. Board meetings, chair meetings, faculty meetings... I still remember laughing aloud as I learned one day there was a "Committee on Committees." Ironically, my time was now so mired in the granular details of *running* a research center, I had no time to conduct any actual *research*. In short, the bureaucracy was killing me.

MAKING WAVES

Just three months in, it hit me: I knew that if things continued the way they were headed, I would never find the time or drive to produce the caliber of work I was capable of. I was convinced I could save lives, and I was also convinced I couldn't do it in my current situation. So, I decided to quit. But the next morning, in a moment of clarity, I revised my plan; I decided to get fired.

What if I just stopped doing the things that didn't give me energy? What would happen if I simply said, "Your committee work is no longer a priority for me, so I will no longer be doing it"? A bold and, frankly, stupid thing to do if you like your job, I realize now. But I had already left this job in theory, so I was free regarding the manner in which I put my decision to practice.

As a scientist, I'm categorically prone to list-making—something about neat and orderly columns just soothes me. So, fully expecting my plan would get me fired, I took the better part of a week to come up with two lists. On the first, I wrote all the things about my job that fit my overall life vision. Things that were actively helping to push me towards my goals. Things that were fun, meaningful and challenging. Honestly, it was a pretty awesome list. On the second list, I did the opposite. I listed everything in my job that was distracting me, or worse yet, directly preventing me from achieving my vision. It was that simple: two lists.

Once I had it all laid out in front of me, I was mildly shocked and disgusted at the size of my second list. It was obvious I would have to make some serious changes if that life vision were ever to become a reality. By the same token, I also intuitively understood that no one can get away with doing absolutely none of the things they don't like to do. So, I compromised at a nice, even ten percent...a number that has since become a rule and tool. If I can work seriously on the things that actively inspire and galvanize my vision, I will put up with ten percent of the administrative *bleh* on that undesirable list.

Generating those lists took me four full days, but this exercise allowed me to visualize comprehensively the job that would empower me. Since I was on a hot streak, I decided to take two more days to untangle myself from all the previous commitments that were eating away at my time.

I called committee chairpersons and gently informed them I would no longer be serving on their committees. I called societies and told them I would no longer be serving on their boards or in administrative capacities. I called professors and told them I would no longer be teaching their classes, grading their exams, etc. For two days, I verbally jiu-jitsued my way through an army of red tape—and upset a lot of people.

At the end of the second day, I was called into the office of the chairman of my department, the man who had hired me. With thinly veiled restraint, he informed me about the litany of angry phone calls he'd been receiving from people all across the university. Needless to say, he wasn't pleased with my explanation of the ten percent rule. Actually, his exact words were something like, "That is absolutely not done! Certain duties are expected of a new professor...without exception! You cannot..." I will never forget the moment that he paused for a breath when I politely reminded him that he could always fire me. We did not part well. I remember a very red-faced man yelling after me as I respectfully bowed out the door.

The next morning and every morning thereafter for months, I looked for that letter of dismissal. It never came. And a very strange thing happened: I noticed unearned respect from people all across the university. I suppose they assumed I must be quite important to have been excused from so many responsibilities. In truth, I had simply set the parameters within which I was willing to work. I had put my foot down for balance.

CONTROLLING THE "MONKEY MIND"

Here's an interesting fact: the brain, unlike any other organ in the human body, can imagine situations that are *hypothetically* stressful and then react to this imagined stress with the production of very *real* hormones and neurotransmitters. So, even if your body is physically in balance (e.g., you eat well, exercise regularly, keep a consistent sleep schedule, etc.), the power of destructive thoughts can override your physical well-being.

As a scientist, a CEO, an entrepreneur and a father, my days are abounded with stress-inducing situations. Financial or legal crises, mishaps in the lab, employee disputes... there seems to be no end to the mental triggers that threaten the stability of those in leadership positions. Left unchecked, this type of stress can cause memory impairment, affect the quality of your decision making, and even negatively impact your communicative abilities. Moreover, in a business (or research facility), the behavioral side effects of a leader's stress don't merely put that person out of balance—they put the entire organization out of balance.

So how best to manage your own mental narrative?

Myriad self-help books, meditation techniques, and martial arts sequences can help teach you how to rein in your monkey mind. Personally, I gravitate to the phenomenon of "picking hands." To illustrate this concept, imagine a farmer picking fruit over a large tract of land—basket hoisted around his hips, body bent under the hot sun, hands moving in the same rhythm for hours. You might think such an activity would cause the brain to become very active and thus generate even more stress as it hurries to complete each subtask. However, MRI scans of brain wave activity have actually shown that people in this flow state are in an energetically inexpensive state of function.

This example resonates with me because I experienced it so often in my own childhood. Growing up on a farm, we rotated our crop cultivation depending on the season. As a result, many of the tasks and chores I performed every day were mind-numbingly repetitive, but also quite complex physically. Unbeknownst to me at the time, I was getting schooled in a very important life lesson: the power of flow. Of course, at the time, it just seemed like a lot of shoveling cow s—t and picking corn. But, sure enough, this state of flow would find me again in a very different iteration of my life—and, man, did it come in handy!

I had a transformative experience early on, in the decades when I had been practicing martial arts—one that highlighted the pervasive value of remaining calm and centered in the face of chaotic situations. In this particular instance, the chaotic situation was a 250-pound, six-and-a-half foot tall, brolic German man with whom I was paired up with to spar one evening in class. To say I was intimidated would be an understatement. The man looked like he ate cinder blocks for breakfast. And besides, he had a higher rank than me.

As the session began, this pile of a human muscle turned into a tornado of swirling limbs. My anxiety spiked. I could sense my movements and reaction times were off and stress was getting the better of me. I felt rigid and stiff, unable to keep my regular pace.

Suddenly, in the middle of the match, I had the sensation of waking up after a bad dream. It was as if I had been on autopilot cruising through a mild nightmare and then, there I was, awake once more. And here was the interesting part: the man saw it too. I watched with curiosity as his face shifted from aggression and confidence to quietly blossoming fear. His movements shifted. For the first time in the fight, he began backpedaling when I came forward, his aggressive stance crumbling into a defensive retreat.

During a high-octane situation, my mind became a peaceful, still pool of water. I was only peripherally aware of the sweat pouring down my face, the frantic heartbeat beneath my uniform. Even while my limbs flailed wildly around like a badass octopus, landing punches and kicks from all different angles, my mind was calm and still. I was light, excited, and happy. It was the first time I had reached what I have come to know as an extraordinary manifestation of martial arts practice—flow.

It's not just Shaolin monks and Indian gurus who can learn to utilize and master this flow state of mind. Translated to business, the ability to find peace amongst chaotic or unsettling stimuli is a very powerful skill. Very often, CEOs have to face their own metaphorical six-foot German warriors (I know because my board room is often full of them). Being able to hold a goal in your mind as you deal with myriad obstacles allows the minutiae and complexity of competing interruptions to become devalued. When you train your mind to turn away from distractions, they quickly lose their ability to exhaust and deplete you of your energy.

When I can take that balance and clarity into my work, it allows me to achieve tenfold results in my performance. I can also tell you with certainty that this skill is infectious. If you leave a super smart team of people in a room with a complex problem to solve, it often results in a mountain of possibilities, each subdivided into a million potentialities...which ultimately leads to no actionable plan. However, if you emulate balance and pull your staff through the sea of dizzying details toward the larger issue, they will mimic your behavior. They will learn to look toward the goal, and suddenly, their challenges will become surmountable and their paths clear.

During negotiations, holding the goal in mind is paramount as you consider several dozen deal terms. But it is equally important to instill that goal into the person on the other side of the table. If you are successful in your mission of placing your vision into their brain, they will let go of the minutiae. If you are in due diligence mode, you would do well to remember this. Details fall aside in the face of awe.

A HEALTHY BRAIN MAKES FOR A HEALTHY LIFE

Ready for an unpopular opinion? Running a business makes the brain old. I know, I know... it's not what anyone wants to hear, but it's true. A body that is consistently parked at a desk or subjugated to repetitive meetings every day is a body that, I assure you, will see a decrease in blood circulation and an uptick in stress hormones. This lack of physical exertion directly impacts the brain by decreasing neuroplasticity—the ability of the brain to wrestle with and learn new things.

Remember when you were told our brain stops developing in our youth? Well, that's not just a little inaccurate, it's flat out wrong. Our brains can continue creating new neural connections throughout our lives. This ability to adapt—what is meant by the term *neuroplasticity*—is quite literally the superpower of the human race. But it takes an agile brain to adapt, and for that, there's no substitute for exercise.

Exercise keeps the brain young, it clears harmful metabolites and hormones, releases brain-derived neurotrophic factors that result from muscle activity, and directly enhances brain plasticity. In layman's terms, there can be no balance where there is no exercise.

Even more exciting, recent science has shown that exercise also increases the general excitability of your brain through a region called the reticular formation. Essentially, while in a state of arousal or excitement, your brain is in learning mode. In this heightened state, your physical senses become sharpened, allowing your brain to work at a much more efficient rate. This is exactly the state of mind a leader needs to be in if he or she is to make effective decisions on the fly. I've never met a single leader who can be optimally effective without exercise.

Leaving the complicated science out of the conversation, many people already know this: exercise is good, sitting down all day is bad. Sure, we *know* it, but how many of us *live* this information in our day-to-day lives? How many of us actually practice balance? All too often, I hear the same rebuttal: time. I don't have time to exercise. I don't have time to meditate or practice deep breathing. I don't have time to eat well. I don't doubt the validity of these people's convictions. They truly *believe* they don't have time, but I would wager it's not time keeping them from living a healthier, more active lifestyle—it's optimization. It's living a life by design, not by circumstance. It's being the machine, not the pinball. To put this into perspective, try asking yourself a few questions:

When in the day do I feel the most energetic?

When do I come up with my best ideas or solutions? At night, right before bed? Over that first sip of coffee in the morning?

When do I most often make mistakes, or just don't care as much?

For most of us, we've probably never thought to ask these questions before. The irony, of course, is that there is simply no way to remain in a balanced state if you don't know what your naturally balanced state looks like! In science, we call this your diurnal pattern and it directly affects when your energy is at its peak and when it's bottomed out for the day. Once you begin to pay attention to your own rhythm, you can design a schedule that suits your energy levels, optimized for that portion of your day.

Then comes compartmentalization.

The key is to group similar tasks together so they segue without interruption or impediment, so you spend less energy shifting between disparate subjects. For example, I schedule all of my meetings that I know will require me to make complex intellectual decisions in the mornings. Why? Because my body tells me to! I know that I personally do my best and most rational thinking at the beginning of the day. Thus, anything that requires my full brain power is clustered back-to-back in the morning.

By the time the afternoon rolls around, I'm mentally burned out. My brain's a little zapped. At this point, I've used many different mental faculties to make decisions, and now my brain wants nothing more than to kick back and cool it for a while on autopilot (think of "picking hands").

So, I give my body what it needs for optimization and head to the gym. Yes, smack dab in the middle of the workday. By scheduling my activities like this, I'm able to exercise consistently four to five times a week. My tired body then demands rest, so I fall into a series of meetings in comfortable chairs, during which I cluster absorption tasks like taking in updates, getting opinions, watching, or listening to my team. It's amazing how much more time you can get out of your day when you group like-minded activities.

Humor me for a moment: as a child I bet you didn't like to eat dessert and vegetables at the same time, right? If I asked you why, aside from the obvious—because you're not a psychopath—you'd probably tell me it's because the flavors don't complement each other. They just don't mix well. Well, the same is true of our daily responsibilities.

Every time I have to rip myself mentally from a serious life-or-death conversation and jump headlong into a meeting about marketing practices and campaigns, there is a definite incubation period. The two don't mix. My mind needs time and energy to shift gears and readjust its state of being—and that's time and energy wasted.

A fundamental practice of balance is not just making sure you check all the boxes, it's checking the *right* boxes at the right *time—for you*!

WHOSE COMPASS ARE YOU LOOKING AT?

I have a friend who is the world's largest supplier of garbage bag twist ties.

He came upon that enterprise after hiking through China, as he and a friend stumbled across a factory town. While in a very remote part of the country, he noticed a series of large towns that were each built around a single industry. All the residents of the town worked at the major factory, their kids grew up to work in the factory, and so on.

In one particular town, my friend decided to tour one of these facilities. It happened to be the producer of garbage bag twist ties—little cards of those green, paper-coated wires that help keep bags closed. Long story short, he crunched some numbers, made some calls—and now he owns the majority of the world's market for garbage bag twist ties.

It's a simple company with very low overhead. He has one employee who tracks orders and transactions on a computer. His business serves as an intermediary between the buyers of these twist ties and the Chinese factory that fulfills the orders and shipments. He's been pulling down several million dollars per year in revenue since his mid-twenties.

He works, on average, a hefty three hours a week.

Sometimes I look at him and think, "Damn, I really did pick a hard way to make money in the world." But I also know I would suffer if I had my life set up like his. My friend is undeniably a rock-star of a businessman, his career brings in loads of money, and he accomplishes a great deal of his Life's mission with that money. But it lacks the sense of mission that I so strive for.

A few years ago, I met with an investor when my company direly needed funding for its cancer and skin care programs. There was a lot riding on this meeting, and it was important I be as prepared as possible. I spent weeks designing a beautiful presentation, refining my pitch again and again, and finally delivered to him all the very promising reasons why he should invest.

I nailed it.

The investor was extremely enthusiastic and actually offered to write a check on the spot. "I'm going to fund you," he said, and he meant to the tune of millions and millions of dollars. Then, to my great surprise and regret, he added, "We'll scrap that cancer program and make bank from skin care, young man!" It was as if the entire mission had gone over his head. We used the proceeds from our skin care products to support our cancer program! Seriously, had he heard nothing I'd said?!

I smiled politely, but inside I was thinking to myself, "You don't get it. I'm not taking your money." And I didn't. The man looked at me and saw dollar signs when he should have seen the countless lives he could have helped save.

I am a huge proponent of physical and mental balance; they are the cornerstones for living a happy and sustainable life. However, I think another, equally important component of balance often gets overlooked: values. One of the quickest ways to fall out of balance in your own life is to pursue paths and courses of action that run counter to your individual moral compass. Sadly, I see this happen all too often. Knowingly or unknowingly, out of desperation or plain indifference, people bargain away their moral and ethical ground to make a quick buck.

Tony Hsieh, CEO of Zappos, once said, "Chase the vision, not the money."

I believed in the statement when I heard it for the first time, and I still believe it now. Money may be the single biggest unbalancer of leaders the world over. This is particularly true for business professionals and entrepreneurs. If the focus becomes financial to the point where the mission and values are lost, rest assured, balance is also going out the window.

As a CEO, I understand that capitalism plus a value-accreting company equals power. Money is certainly a brick in that build. But building and maintaining power can only be done by a leader who knows balance and can keep an unfailing eye on the mission.

Fear not; money will follow when you enact positive change in the world.

To sustain this type of business model and keep myself in balance, I actively seek to ally myself with people who have a grand vision—one that encompasses more than just a big payout. It's not altruism—far from it, actually. It's self-preservation. If I had accepted that investor's money way back when, I would have put the entire company in jeopardy: the organization would have lost its balance, and as a team, we would have lost our sense of purpose.

And yet, it would have been so simple. All I had to say was "yes."

Comfort leads to complacency, which ultimately leads to boredom. Put another way, comfort can throw you out of balance as a leader. As human beings, we are incredible adapters, and once we have ourselves in a comfortable environment, there is a tendency to do whatever it takes to maintain that level of ease. When leading a business or organization, it's easy to find yourself shifting toward an attitude of maintaining minutiae: carelessly making decisions to avoid difficult or challenging situations. Complacency kills, and I'm not just talking about profits. Know your principles and stick to them. In business. In your relationships. In the way you treat others.

Balance truly is the other everything.

So, make your list. Find out what you love, and what you're willing to tolerate to get to that place. If this is the only thing you do to invite balance into your life, you will see a difference. I promise.

5—LEADING WITH BALANCE

The Canadian journalist Malcom Gladwell wrote, "The visionary starts with a clean sheet of paper and reimagines the world."[1] In my field, the world-building has been ongoing for thirty years, and there is hardly a stick figure on the page. You know the thing about changing the world? It's hard.

VISIONARIES ARE JUST PRODUCTIVE PEOPLE WITH OCD

As a new assistant professor at the University of California, Irvine, I was massively disheartened by the lack of progress in the field of spinal cord injury treatment.

---

[1] *Gladwell, Malcolm, Jill Lepore, and Malcolm Ross. "The Real Genius of Steve Jobs." The New Yorker, November 7, 2011. https://www.newyorker.com/magazine/2011/11/14/the-tweaker.*

Decades of intense study and research had turned up almost no tangible results. We were still a long way off from successfully repairing the nervous system of a common field mouse, let alone a human being. I was staring at the blank infinity of an unmarked page, and it was staring right back at me, daring me to make a move.

While I might have a higher IQ than your average potato, I realized that if thirty years of research hadn't turned up an answer, my intelligence would not be the X-factor that would save the day. I needed to look at the problem differently. There was something out there that none of us was seeing.

Enter the visionary.

The thing about people with vision is, more often than not, they tend to be a little... *off kilter*. As embarrassing as it may be to admit—and recognizing the inaccuracy inherent in the self-diagnosis of mental disorders—I consider myself a "managing hypomanic." From a practical standpoint, this makes sense. To get the kind of results the world has never seen before, to take a stance that's never been taken before, you need to be okay with (or unaware of) the fact that you are different. Think Steve Jobs or Mark Zuckerberg—not exactly your high school prom kings. However, what these two had in common was an innate ability to remain hyper-focused while living in a tumultuous sea of granular detail. Visionaries are like pit bulls with lockjaw; once they sink their teeth into something, it's almost impossible to let go.

During my time at the University of California, Irvine, I happily joined such a tribe of misfits. For ten to sixteen hours a day, I stared down the lens of a microscope looking at injured spinal cords. At night, I lay in bed and dreamed about spinal cords— not tropical beaches or yachts on luxurious islands, but *spinal cords*. I thought about them all the time, lived with them like a put-out in-law. For over a year, I allowed myself to be consumed by the need to find a new therapeutic treatment. I had cast aside all prior work and preconceptions and set off into a medical no-man's-land, determined to come back with something revolutionary or nothing at all. I knew this path could very well lead me to nothing, no grants or publications, which would cause me to lose my new professorship. The safe thing for a new professor to do is to follow on the heels of the great professors around you. But I couldn't do that; their paths had not yielded treatments, and I needed to find a new path.

After a year of obsessive-compulsive research—a very uncomfortable place to reside, I might add—I had a life-changing breakthrough. I'd found a new therapeutic target. That was the beginning of what remains to this day the most advanced spinal cord injury treatment in the world. Thank you, OCD.

I believe visionaries are the driving force behind creative disruption. They are the game changers who come along and rile us up, the ones who challenge stagnant notions of the status-quo. Yet, for all that, visionaries are only as good as their counterparts.

The most effective visionaries and CEOs lead with balance to their counterparts. I've become acutely aware of this truth, and I am blessed to be surrounded by a team that grounds me. There are outstanding people around me every single day, many of whom have followed me from company to company. These are the individuals I trust to balance me. They will listen with interest and immense patience as I talk excitedly about a new concept I've come up with, or the fledgling of a great idea I think we should implement. And then one or another—I sometimes think they take turns—will raise a hand and say something like, "Hans, that's a brilliant idea. It's just that you have no idea how much work that will take."

I encourage this reality check from my team as it tethers me to the practical and the pragmatic. Ultimately, it keeps my ego healthily in check. Unfortunately, I've witnessed many great visionaries and CEOs come undone after they began drinking their own Kool-Aid. By putting too much stock in their own perceived brilliance and too little value on the hard work of a team, they lose sight of the original vision. It's a form of hubris and a slippery slope toward imbalance. Sometimes, this disastrous overreaching can even result from good intentions.

My friend Mark is a good example.

Growing up, Mark was fanatical about VW Beetles. To him, they were a well-polished symbol of success. The veritable American dream incarnate. Thus, it wasn't surprising to me that, as he started his climb up the corporate ladder, he bought one for himself. What did come as a shock, though, was the day I went with him to visit one of his offices and noticed a parking lot full of VW Beetles. All of his employees owned one, all the same make and model. "What in the...?" I thought to myself.

"Umm, Mark...why does everyone have the same car as you?" I asked.

With a mischievous smile, he looked at me and said, "I gave each employee a car as a perk. I lease them on the company dime." Waving an arm at the parking lot, he added, "This is what success looks like, my friend."

I didn't say anything at the time, but I remember thinking how just six months prior he had been desperately trying to raise more capital. The company had been almost bankrupt, and he was facing the possibility of a total shutdown. At the last minute, he had corralled an investor to unload $250,000 into the company, enough to keep things afloat. In all his wisdom, he invested this influx of capital into eight shiny, new cars. As you likely suspect, his company died within a year.

This was a huge lesson for me about how a CEO, or any visionary, for that matter, can throw a company out of balance. Vision is a powerful tool, and one I think is absolutely integral to any concept of innovation. However, with vision must come balance. Lofty ideas and good intentions are only worth the foundation on which they are built.

CHARACTER OVER CREDENTIALS

My first interview with Garry took him completely by surprise.

I started with my standard routine of interview questions and small talk to get the candidate talking. Midway through the interview, I decided to shake things up by inviting him for a walk to a nearby coffee shop as we continued our conversation. I tried to make the walk a casual, safe, and comfortable space for Garry to open up and tell me more about himself and his life in an informal way. As we danced in circles throughout our conversation, I couldn't help but notice there was a two-year gap in his narrative. Two years that he couldn't, or didn't want to, explain.

I always look for gaps in these interview conversations, as they're typically red flags.

I tried to approach the subject of his missing work history three or four times, but he politely refused to go there. As we were walking back from the coffee shop, in the middle of an otherwise very pleasant conversation, I told him bluntly, "Well, Garry, I'm sorry but you don't get the job."

He was shocked. He stopped dead in his tracks and said, "What? I thought this was going well?"

"It's been fantastic," I told him. "It's just that I need a right-hand man. I need somebody I can talk to, and more importantly, believe in. You're just not giving me that. I can't trust you if you're holding things back from me. I don't need a yes-man; I need somebody who can stand up and challenge me and the system and think laterally. I need somebody who's got some honesty and bravado, and you are keeping something from me. I just can't have that if you're my right-hand person. I need to know we have a completely open line of communication."

His shoulders slumped. He looked at his feet and was silent for a minute. Then he told me his story, a narrative he'd never even hinted at before.

Garry is Hungarian. He and his family came to America to escape a dire situation in Hungary. Unfortunately, Garry wasn't licensed to practice medicine in the U.S., so he and his family started their lives from scratch here. At the time, he didn't speak English and work options were scarce. So, he taught himself graphic design and worked in an illegal sweatshop for 12 or more hours a day.

When I asked him why he chose to do that kind of job, Garry responded, "I couldn't practice medicine anymore, and I had to feed my family." With pride, he added, "It was work, and I would've taken any kind of work at that point."

I hired him on the spot—not for his credentials, but for his character.

What I saw when I looked at Garry was an exceptionally hard-working and dedicated individual, willing to do whatever it took to get the job done. His story is a perfect illustration of lateral thinking. He was willing to shake up his life for his principles, and that is exactly the kind of person I want to work with every day.

Garry has continued to inspire me for over twenty years. I was privileged to watch as he bought his first car, then his first home, and then a home for his parents, whom he had brought to the United States from Hungary. Aside from the strictly monetary side of things, Garry has the freedom to do work for which he is highly qualified in an environment of immense professional respect. It's been wonderful to watch him grow into and utilize his full potential. I love that the companies I build can help change the lives of the people I employ in such a deep and fulfilling way.

None of this would have been possible, however, without the "dude interview."

The people who apply to work within my companies are typically highly qualified. They have years of training and have passed rigorous selection programs throughout their years in academia. Many of them have completed PhDs and medical degrees. On paper, nearly all of them look like exceptionally qualified candidates. In reality, having a prestigious degree from a reputable university only tells me the individual is good at learning new information. But it doesn't tell me a single thing about their character, likeability, or capacity for empathy and teamwork.

Good interpersonal communication skills are fundamental to any business operation, but they are especially critical in startups and fledgling companies, where the reliance on team members and the potential for failure are much higher. As a CEO, it is my responsibility to ensure there isn't any deadweight, that every person is nothing short of enabling to our mission. And equally important, I can't have any glitches in the company culture.

Making hiring decisions based solely on the qualifications of the applicants would be a surefire way to throw my company out of balance. Instead, I rely on what I call the "dude interview." By side-stepping the traditional interview questions, I choose to focus instead on a character evaluation. My questions during the "dude interview" are very informal, highly personal, and intended to keep the candidate talking. I'm looking to gauge my own internal barometer about how this person makes me feel and what kind of reactions they're inspiring in me.

The fact is, I can teach almost anyone how to cultivate stem cells in a lab or perform business functions at my side, given enough time and resources. It's much, much more difficult to train someone to become emotionally intelligent. For me, character always outweighs credentials. I would much rather have a room of highly motivated individuals with no relevant skills than a reticent group of trained professionals.

A fundamental principle of leading with balance is making hiring decisions that continue to foster the type of moral values with which you want to surround yourself.

DON'T SUGARCOAT THE BAD STUFF

There is a pervasive and destructive myth among CEOs that, when there's bad news in the executive suite, it's best to keep it outside the realm of public knowledge. Rather than being upfront with their employees, many leaders choose to go in a different direction. Relying on hollow platitudes or catechisms, they assure their staff everything is just fine. Nothing to worry about here.

This type of communication inevitably throws things out of balance.

For starters, employees have the same psychic ability as new moms—they can just sense when something is off. When that happens and you choose not to address the elephant in the room, you inadvertently get the rumor mill working overtime. Employees who aren't told the truth will create their own narrative to explain the odd behavior of their upper management—and their explanations are often much worse than the actual problem. Finally, not being straight with your employees engenders distrust and ultimately demonstrates a lack of respect. If you trusted them, you'd know they have the emotional fortitude and skill to handle any challenges you might face *as a team.*

You just can't sugarcoat the bad stuff.

I sold California Stem Cell, a company I founded, to a publicly traded entity and was obligated to work for them for 18 months. The sale was a success, but the aftermath turned surprisingly grim. During my 18-month lock-up, they ran into some serious financial troubles when one of their prior assets failed, so I decided to buy back all that I had sold them. But they were in a death spiral and concerned about their salaries, so they shut down most programs and laid off every one of my former employees and many of their own.

It was bad. So, I communicated. I gathered my former employees and told them point blank: "Team, you are being laid off by our new owners. I need three months. Go get another job if you want, but I've managed to get you three months of full severance pay, and I will try to buy our assets back and offer each of you jobs during that time."

I followed through on this commitment. It did take me three months, and some people did take other jobs, but my core team stuck with me through the hard stuff (or, perhaps in some cases, lounged on a beach for three months while I dealt with the hard stuff). The very first day of that three-month period, when I had no company and no money to pay him, my right-hand scientist Gabriel showed up and worked, keeping all our cell cultures alive and experiments going. He didn't just come that day; he came every single day, seven days a week, for the whole three months.

He never once asked about salary or compensation. He just said, 'Hans, I trust you. You're going to pull this off."

I didn't dance around the truth, but told him honestly, "I think I can pull this off, but... I don't know for certain that I can."

"Yeah, yeah, yeah," is all he said, "You do your thing, I'm going to do mine." Gabriel is a phenomenal person, but he's not the only one in my company who has stuck with the mission during tough times.

Twice now, in the early days of one startup or another, I've had to call my staff together and say, "We don't have the money to pay you next month. At the end of this month, you'll all get your paychecks, but I don't have the money to make payroll the following month. If you don't want to show up for work, you don't have to. I won't hold it against you. I'll even help you get a new job; I'll write you a great letter of recommendation, whatever you need. But, if you're willing to hang in there with me, I'm confident I'm going to raise the money for our mission, and you won't see an interruption in your pay. But you must understand that I might not be right. Seven out of ten businesses fail, and ours might be one of the seven. But for the sake of our mission, I am going to hit this with everything I have."

On both occasions, every single employee stayed with me. What's more, both times I was able to raise the money, and everyone was paid on time. When I asked my people why they stayed despite the financial risk, they all said the same thing: because of the way I treat them. They felt respected, and as a result, they trusted me. They kept their eye on the goal and trusted me to do my job. The outcome? Mutual trust and a renewed dedication to the mission.

I lead with balance by treating my people with respect, by telling them the truth (even when it's uncomfortable), and by following through on my commitments. But honesty and transparency are a two-way street.

In business—especially biotechnology—programs often take far more time and money than anticipated, and one must have integrity to own the slip in time or the over-expenditure of resources. Thus, integrity becomes a trait that is not only necessary for a CEO, but essential for the entire team to function cohesively.

I impress upon my staff that a miss or a failure is never a bad thing, as long as they tell their colleagues and me that the miss is coming before it happens. If they don't inform me of their slip and I find out later, then ultimately, they have failed. If, however, they tell me in advance that they're not going to hit their deadline or they're not going to get that bit of data, they're viewed as a worker with integrity. This behavior is something that's extremely important to highlight with your greater team. Every time this happens, I publicly praise that individual for having the insight and diligence to foresee a problem. And by turning your own anticipated slip or failure into a demonstration of integrity, you lead by example and create a culture and a communicative environment that encourages balanced leadership.

Don't keep your employees in the dark. Everything good starts with honest and open communication.

YOU'LL NEED A FORMIDABLE SENSE OF HUMOR AND HELP

Day in and day out, I'm surrounded by death and disease. Given the nature of my medical research, I am routinely faced with decisions that end in life or death. Families are torn apart, people are in agonizing pain... and the truth is, we can't save all of them. That's a hard thing to swallow, no matter what profession you're in.

I've got heavy all around me. I've got heavy in spades.

Part of leading with balance is developing an unshakeable sense of humor, a light-ness and gratitude towards the world. Without this characteristic, the stress of the job will wear you down to nothing. All too often, I see CEOs and other high-level professionals become the bottleneck in their own systems. Sometimes it's an inability to relinquish control; other times, it is a lack of counsel, or surrounding your-self with the wrong counsel. The fact is, we even encourage and celebrate these arche-types of the silent and stoic leader. But the result is always the same: without humor, without gratitude, you will always get stag-nation, decline, and crisis.

As an individual in a leadership position, there is simply no way to avoid stress; it's an inevitable component of the job. What I have learned over the years, however, is the importance of learning to look on the lighter side of things. For every patient I lose, there are many others who survive. There is someone out there who gets to start a family now and make their dreams come true. There are more miracles happening around me every day than tragedies, and it's important to take notice of both. Whether you choose to use a gratitude journal, a special mantra, or simply a prayer, I cannot stress enough the value of consciously cultivating a light heart. And I'd be lying if I said I could always do this on my own: I rely immensely on my executive assistant.

The important qualities of a helping hand (e.g., an executive assistant) are not so much skills or tools. Booking flights, managing schedules, greeting people at the door—all that stuff accounts for about five to ten percent of the reason I might hire someone in this role. The other ninety to ninety-five percent is all about personality. When I'm thinking about selecting my executive assistant, my personal go-to employee, I barely consider their clerical skills. I need someone around me to balance out the heavy; I need someone with bright light and energy, someone who will remind me of the mission on the days I feel overwhelmed by the sadness of the diseases we are trying to fight.

A good executive assistant should act as a foil.

Take a law firm or an engineering company, for instance. In both of these examples, you can expect the company culture to be fairly homogenous in terms of composition: lots of mathematical, type-A people. Which is great...that's what you want for that specific purpose. But your executive assistant, your helping hand? Not a chance. If you choose someone like that, someone who is a mirror image of yourself and everyone else in the company, the only feedback you're going to receive is a Xerox copy of what you've already said. Essentially, you've done nothing more than adopt a parrot. The foil to this type of personality would be someone who is light-hearted, energetic, and enthusiastic. Balance is about finding the complement to the frame of mind you wear most, not piling on more of the same.

When I consider what it means to lead with balance, for me, it is about so much more than just the mechanics and metrics of getting my job done. It's about bringing the right attitude, it's about equipping myself with the skill sets and emotional coping mechanisms to lead and inspire others, even in the face of tremendous hardship—and I can't do that alone. Leading with balance might start with you, but it requires something else on the other end of that see-saw: all the people who depend on you.

6—PUT THE RIGHT THINGS IN FRONT

Stan Lee once said, "With great power comes great responsibility." While I am aware that Spider-Man and Uncle Ben are fictitious Hollywood creations, I'd ask if that negates the validity of the message? I'm inclined to say no, it doesn't. The potential to take something from your mind and manifest it is one of the premier perks of being a CEO. However, with this immense power comes the twin flame of responsibility.

As with Peter Parker's transformation into the amazing Spider-Man, the journey to becoming the hero of your own narrative is fraught with perilous traps—specifically, of the ego variety. When it comes to putting the right things in front, it's about much more than just setting the right priorities for your business. To understand the trajectory of a company, it's imperative to understand the mindset and values of its leaders.

BEWARE OF "BUSINESS CARD SYNDROME"

Business Card Syndrome is a diagnosis I give to executives who place too much emphasis on the trappings of their titles. My first encounter with Business Card Syndrome came about while watching a close friend's journey into entrepreneurship.

To him, it was of paramount importance that he be able to print a business card with the neat, little letters C-E-O trailing his name. He loved handing that card out to people; it was a source of immense pride and joy for him. While I love him dearly, his desire to prove himself as a *man of status* ultimately drove him to put the business in a secondary or even tertiary position to himself—as exemplified in the earlier story of a parking lot full of VW Beetles.

Business Card Syndrome is particularly dangerous because it leaves a company's leaders vulnerable to incompetence. The reason being, if you're intensely focused on acquiring a specific title or status, by the time you reach said position, you are almost certainly going to be thin on the skills required to do the job. Until that point, after all, you've just been chasing a title. A title doesn't require you to learn new skills or overcome your weak points as a leader. Contrarily, it allows you to get away with cutting corners and even seems to reward the single-minded focus of the maniacal.

On the other end of the spectrum, we find leaders who are motivated and galvanized by the mission, that is, *the company's mission.* Mission-driven leaders are never thin on the skills required to execute their positions effectively. For these individuals, the idea of a title comes second. If anything, a title is a natural byproduct of a more important goal: the success of the company's mission.

When it comes to leadership, I try to keep a healthy distance from anyone who gets overly excited about three-letter acronyms.

To illustrate this point...

Not too long ago, I parted ways with an employee who had been my vice president of business development. For eighteen months, this guy killed it. He was nothing short of phenomenal at his job. Yet, in those eighteen months, on three different occasions, he approached me and said, "I would like to be promoted to Chief Development Officer."

The first time he brought it up, I said, "Well, you know a CDO is someone who brings in deals. They have their own list of contacts that they bring to the company, and they actively close new deals with said contacts. Let's take a look: in the last six months, I brought in all the new deals. I brought in all the new money. You were an incredibly important asset in helping me take care of the analysis and paperwork, but you didn't source anything."

Twice more he came back to me with the same request, and twice more I asked him if he had sourced anything new for the company. It was an awkward little dance we were doing, and it gradually became evident to me that the title Chief Development Officer was more important to him than the actual work. Eventually, as I had anticipated, he turned in his resignation. Fortunately for him, he went to work for a company that offered him the title he desperately wanted. Not so fortunately, he ended up getting laid off soon thereafter. He didn't yet have the skills to do the job effectively.

It was a classic case of Business Card Syndrome in action.

ONE THING FROM WHICH EVERYTHING FOLLOWS IN ITS WAKE

Among other faults, an egotistical CEO is likely to compromise their fiduciary responsibilities. I've been privy to many occasions when a founder or CEO abused his or her position by misusing investors' cash or not putting in the work to achieve the proposed missions and goals. When you build on the wrong foundation like this, it's nearly impossible to get something that will stand on its own legs. You commit yourself to failure.

Putting the right thing in front, from a CEO's perspective, is simple—at least to describe. In every complex issue or undertaking, there is one thing from which everything follows in its wake. In biomedical or scientific endeavors, for example, that everything is medical or scientific excellence.

At one point in my career, I was an active member of the Board of Directors for a stem cell company. Gradually, I saw the science starting to fail. This, in and of itself, was not a cause for alarm; when it comes to scientific endeavors, many of them fail. Yet, instead of changing course or adopting another project (i.e., maintaining a commitment to scientific excellence), the company leaders put science on the back burner and placed money in front.

They began to exaggerate the test results and even resorted to flat-out mistruths in order to overcome the deficiencies of the science. They were compelled to impress the investors at any cost. And it worked for a time. Their salaries climbed to heights that became more and more difficult to sustain.

Meanwhile, I decided to vacate my position on the board. I was not willing to be part of a firm whose moral imperatives allowed them to lie if it meant a pay raise.

I had an employee once whose job it was to interface with the FDA. She was my only employee in that role, as I didn't have the funds to hire a full internal regulatory team at the time. Instead, I hired consultants who came and did their work, collected a paycheck, and left. My employee collaborated with these consultants and then filled in the gaps to put together what was needed to make our final FDA submissions. I knew it was extremely challenging for her because she was in an underling position with the consultants, while being my most senior full-time regulatory employee.

At one point, she came to me with tears in her eyes and said, "Hans, I so appreciate you teaching me, bringing me up in my career, and giving me these positions of increased responsibility, but I think I need to work for a large company—one where I can report to a full-time senior staff from whom I can learn more." I understood completely, and while I was reluctant to let her go, I was happy to see her bounce from my start-up to a major corporation where she could grow.

Three months later, she reached out to ask for her job back.

"Why?" I asked. "You've got everything you wanted. You're in your dream company now."

"Yes, I'm in this huge company with highly experienced colleagues all around me," she said. "Five thousand employees under one roof, and every single one of us knows the science is destined to fail. I've been dedicating my life to help aggrandize a pseudo-scientist with a flawed medical dream."

The company was run by a billionaire who had been a successful professor. After amassing great wealth through his work, he single-handedly financed the company's operations, thus rendering a board of advisers and sophisticated investment firms unnecessary—how convenient. Unfortunately, the man had a massive ego, and it blinded him to the shortcomings of his business and scientific pursuits. The company mission and directive were out of whack; rather than delivering a quality product, the goal became saving his professional reputation at all costs.

It's been over ten years now and the company has remained stagnant, angrily limping along thanks to the enormous financial contributions of its founder.

An ego that is front and center will always dwarf the company's mission.

A leader who disproportionately dedicates time and energy to fancy business cards, the company address, and a beautifully furnished office is likely putting ego in front. I want to be clear here: there's nothing wrong with having an elegant business card or an office with a putting green. I'm all for that. However, when that becomes the main focus, when it becomes a way for that CEO to reinforce his or her own positive identity... you're in for a whole lot of trouble.

KILLING THEM WITH KINDNESS STILL LEAVES 'EM DEAD

The ego works in mysterious ways, and it's true that the road to Hell is often paved with good intentions. While some business leaders fall in love with their own legacy and actively distort the success of a company for their personal benefit, others do so almost ...accidentally.

A subtle form of CEO ego-preening can seem counterintuitive: extreme generosity. Returning to my friend and his fleet of shiny new Beetles, he was very open-handed with his staff, and of course, they loved him for the great pay and all the perks he made available to them. But his company was in no position to sustain that level of expense, and there wasn't a sense of ownership or earned reward among his employees.

When it finally went under, all of those people lost their jobs.

I like to pay people well, but not as the overriding priority in the company. My staff has to work hard, and my employees need to feel a sense of ownership. It's important to me as a leader that they know the rewards they receive are based on merit, not charity. A leader doesn't do anybody any favors by being overly kind.

A while back, I was being interviewed by another entrepreneur as part of an award nomination I had received. The interview began with this entrepreneur giving me an exhaustive list of his credentials—he had built a company that was eventually valued at two-and-a-half billion dollars, he'd won a litany of awards, so on and so forth.

"Wow, that's extremely impressive," I told him. "How is the company doing now?"

"Oh," he said, suddenly bashful. "Well, unfortunately it crashed and burned."

As if making up for that one tiny detail, he quickly told me with great pride that while the company had indeed gone under, during his fifteen years as its figurehead, he had never fired an employee. Not a single person.

"I took responsibility for those people. Once they were in my family, I was committed to them. Unfortunately, 51 percent of my company was bought by a private equity fund and they made me fire people. A lot of those people ganged together and filed a class-action lawsuit that ultimately brought the company down."

I listened attentively, but all I could think was: you lost your company because you never fired anyone, not because of any action of your acquirer.

Adherence to a no-firing policy is just as egotistical and irresponsible as a parking lot full of new cars. I'm sure his under-employed staff provided the CEO with tremendous positive feedback when they halted their video gaming to smile at him as he walked by their desks. If your policy is never to fire anyone, how can you possibly be putting the right things in front? You're not empowering the drivers of your mission, you're actively undermining them by rewarding mediocrity, or worse still, downright incompetence. The values of this entrepreneur, at least from a humanistic perspective, were commendable. In the end, however, his actions ultimately fed his own ego more than it helped the business.

I fire fast, and I tell candidates that upfront. It's not my prerogative to be mean or cold hearted, but it is my responsibility to put the right things in front—and that means having the right people in the right positions. I can't let my personal ego get in the way of doing what's best for the company. To be honest, I hate firing people. It goes against my nature and my livelihood to cause people pain. As a CEO, however, I've learned to do what's necessary for the larger mission; I've learned to separate my self-worth from that of the company.

Altruism, however well intentioned, is a poor business model.

"STOP BEING SO DAMN MODEST!" AND OTHER COUNTERINTUITIVE ASPECTS OF LEADERSHIP

Humility is a beautiful thing. Heck, I was practically raised on it. Growing up on the farm with very little money, I learned the value of humbling myself before others. It was an important part of my life then and one that has ultimately served me well. However, like all aspects of leadership, it has its time and place.

One of the most surprising lessons I've had to learn over the years was the potentially disastrous effects of modesty. Sounds odd, right? Well, it's about to get even odder: there are situations where your modesty doesn't just help you fail, it can actively hurt the well-being of your company.

The business side of a biotech company demands the antithesis of modesty in a CEO. The company's board and employees do not want their CEO to be modest. Ask them to choose between a quiet and deferential monk or an outspoken warrior—it's pretty obvious they're voting for the guy with the axe. They prefer a leader who is bold and conclusive, one who brags about the company and instills confidence. A modest CEO, even one who possesses great personal characteristics and professional accolades, is likely to fail to portray the company's brand image effectively.

When I am pitching to a group of investors or giving a keynote speech to a room full of businesspeople, I must put the right thing in front of that audience—more often than not, the way to be aligned with my mission means actively *not* being modest. The unfortunate truth is, a CEO who is tentative or spends a lot of time speaking about how he or she stumbled upon success does not instill confidence. I know because I've learned this the hard way.

My natural default is to step back and share credit, respecting my colleagues and giving them their time in the spotlight. While some might find that admirable, it doesn't always align with my company's mission. I'm a scientist, an expert in my field, and it is important that I instill confidence and am able to speak to the entirety of the subject matter with authority. While it seems decidedly ironic, I'm not being cheeky when I say it is humility that demands the assertion of confidence. Given my role as CEO, there are times when I must put confidence in front.

Now, before anyone grabs the pomade and adopts a cocky Tony Soprano accent, there is a flip side to this. The technical side of the biotech business sometimes calls for a degree of modesty in the CEO. There are some situations in which the right thing to put in front—again, the attribute that will drive the mission—is modesty.

Here is an example of what I mean. During a presentation I was giving in a convention room of over 2,000 people, I failed to note that my audience was not a room full of CEOs and investors, but rather, a room full of highly experienced scientific thinkers. I was talking about a major accomplishment, one in which my lab and I took human stem cells and transformed them into a high-purity population of cells for the first time in the world. By subtly altering their environment, we were able to program these stem cells to become anything in your body—in this case, a spinal cord progenitor I had shown to cause regeneration of injured spinal cords. It was this very accomplishment that brought me a certain notoriety as a scientist.

During the presentation, I was very outspoken and confident. And although I didn't mean to come across as arrogant, I may have earned that descriptor when I said, "This discovery is a first in the world, the highest purity ever obtained, and it's essential for all these reasons... Look, we made rats walk again, and now we're restoring hand and arm movement and sensation to previously quadriplegic humans."

The thing about giving a talk at the front of a room is that you can see the faces of every person in that room. As I enjoyed my confident monologue, I witnessed a room full of interested, wide-eyed faces slowly turn sour. Frankly, they probably thought I was a pompous ass.

I was speaking to a room full of scientists, a collection of brilliant minds, and most of them had been trying to do what my team and I had just accomplished. They recognized the need, but no one had been able to do it. And here I was up on stage, bragging about it. They felt denigrated and challenged. I saw these brilliant colleagues begin to shut down, to stop listening.

Remember what I said about tossing humility out the window? Well, go find it again.

In this instance, the audience called for a modest leader; it served my mission to present myself in a humbler manner. Sensing the loss of interest and realizing what I had done, I shifted my communication style on the spot. After inserting an uncomfortably long dramatic pause, I began again from a different perspective. "You know," I explained, "My team and I are not particularly talented scientists, but we deeply respect and value the work of our predecessors. This would never have been possible without those before and beside us. We could not have done this without all the work of people like you and the people who have come before you. So if I have any talent—if my lab has any talent—it's our ability to listen to all the amazing scientists around us, the scientists who truly enabled this. Every scientist stands on the shoulders of those who came before."

Boom! The lights came back on. The audience was interested again—and not just interested; they were proud.

I'll never forget that moment—watching the crowd turn away and realizing in an instant how haughty and outspoken I had been. All it took was a change of perspective and a quick dose of modesty, and just like that, everyone was back again. If I had been giving that same talk to a group of investors, they would have resonated with the outspokenness with which I began. They would have expected the warrior. They would have been thinking, "Okay, this is great; I want to buy that. I want to partner with that, I want to fund it."

Put the right thing in front.

If the situation calls for you to be assertive and confident—dare I say, even a bit bombastic—then, by all means, bombast away! Whatever drives the mission is what needs to be in front. If it calls for you to be modest and humble, channel that inner introvert, share the glory. If I am talking strategy with my board members, all veteran strategists, my being outspoken and overly assertive won't work. At best, it will come off as disrespectful and impudent, exactly the way it did with those scientists. My general rule of thumb is this: when like talks to like, use modesty. Put the right thing forward, but *as the context demands.*

Appreciation is another critical component for a CEO to cultivate in order to put the right thing in front. Specifically, I am referring here to appreciating the people in the company who get the work done. It's disturbingly easy for a founder or CEO to put blinders on and start believing he or she is the cornerstone of the business's success. "Me! Me! Me!" the mind screams. "Things are all happening this way because of me!"

Woe to the company that finds itself under the direction of this type of leader.

You might be the visionary, the prophet, or the idea guy, but know this: ideas are cheap. I've got loads of them in my head, and they don't mean anything if they're not manifested. On my own, I couldn't have created anything even close to the companies I have led, and I keep that in mind every single time I walk into my lab.

On the heels of appreciation is the neglected stepchild of leadership characteristics: acquiescence.

In this case, I am referring to the conscious act of giving up ownership.

For a person to say boldly, "I'm going to cure cancer" sounds almost ridiculous. But that's what entrepreneurs do. That's what scientists do. They take on the big challenges, the impossible things. And I'll be the first to admit it—it's hard for people like us to give the credit away. After all, you have to be a little vainglorious to believe you can do something no one's ever done before. But it's a slippery slope from pride to egomania. If I, as the company leader, become over-enamored with my ownership of the treatment, the product, the idea, etc., I rob the mission of its power.

True, the visionary is the one who thought of it. They did the extraordinary build that allowed that vision to manifest, and there's certainly deserved pride there. The reality is, though, by the time that vision is brought to fruition, there are 5, 10, 50, 100, or perhaps even thousands of people who have helped make it real, and nearly all of them are thinking the same thing: "I did it, I made this happen."

I developed a treatment for spinal cord injury and handed it to several doctors to utilize in a clinical trial. Two months later, I listened in shock as one of those doctors told his patients, "I founded this treatment." From the MD's perspective, he was the first doctor in the world to use this treatment, so it belonged to him. The same thing happened when I handed treatments for ulcerative colitis and rheumatoid arthritis to doctors. And again when I handed a treatment for cancer to doctors.

"What arrogance!" I thought when I heard this for the first time. However, upon reflection, I told myself, "Wait a minute. Maybe I'm the arrogant one." That doctor is part of the vision, moving that treatment out into the world. He owns it as much as I do. So do the people in my company who worked on the treatment for thousands of hours. So do the investors who chose to fund our study rather than some other. And so do the patients in the clinical trials who take such a brave step, being among the first to receive a new medical treatment.

Being a miser gets you approximately nowhere. Similarly, hoarding ownership does not further medical or scientific excellence; rather, it debases the pure, undiluted mission of the company by infusing it with one's ego.

Let go. There's enough room for all of us at the table.

NO BUSINESS IS AN ISLAND

Food for thought: how many of your best business relationships came from a cold call? Social media request? Text message?

I'd wager not too many.

The reason is quite simple: people don't trust people they don't know. It's not you per se they don't trust, it's strangers, in general. In fact, there even seems to be an unspoken agreement to *mistrust* anyone who touts their own expertise or who pushes an inkling of an agenda. Nonetheless, making strategic connections is an absolutely essential role for any CEO attempting to put the right thing in front. As a result, the most efficient way to get someone to understand and value your expertise is to have that expertise verified by people close to them.

In other words, it is all about who you know...or, more accurately, who *they* know.

When I seek access to a person of influence whom I don't already know, I use a strategy I devised called the *three-touch system*, based upon the idea that someone will be open to becoming acquainted with me if they have already heard about me from other trusted sources. It works because of two simple principles:

1. We have few degrees of separation from any other person on this planet.

2. Your greatest champion is not you; your most powerful advocates are the people you touch.

The three-touch system is about putting the right things—your most powerful advocates—in front. It's a tool to instill confidence and trust, a long play to establish a relationship with someone new.

The strategy is simple. I contact two people who know the person I want to access and ask those people to mention me to my target person in some way. When that has happened, I reach out to that person as the third touch point. During that initial contact, I don't seek anything from them or try to pitch anything. I have no agenda. I'm looking for a person-to-person connection.

The point is to build a personal perspective. I don't lead with my business. I may focus on the charities with which I'm affiliated, a hobby, or on some commonality in our families...anything that is indicative of personality.

Companies with the greatest technology can fail for many reasons that are unrelated to technology. These contextual failures can be mitigated by a leader who is trustworthy, dedicated, and knowledgeable. Thus, the three-touch system is particularly effective in securing investors. No one knows your business, your technology, or your science like you do; and good investors and partners know they can never grasp the minutiae the way you do. Thus, your primary goal in connecting with investors is to show them you are trustworthy, dedicated, and knowledgeable. These attributes can certainly be conveyed by people who know you, even if they do not completely understand your area of expertise. Every investor or partner who has concluded a deal with me did so *primarily* because of their understanding of me as a person.

Investors do not invest in widgets; they invest in people.

Using the same mechanism, I've also been able to make a difference in government policy.

I once taught a doctor from the other side of the world how to perform microsurgery and grow stem cells in a lab. He came to take a three-week course that my institute provided under my leadership. During our time together, I spent many hours teaching him, talking to him, and getting to know his life and mission. It was not work at all— he was easy to be around and I found his natural mission to help others refreshing and relatable. Several times in the subsequent decade, as part of my ongoing efforts to keep my relationships current, I contacted him to check in and offer help when I could. Later, he became a personal physician to the president of his country and one of his close friends became the minister of health.

During the same timeframe, I came to know a senior U.S. Embassy official to that same country, helping him with several personal matters on a recurring basis. When I wanted to contact the president and the minister of health to discuss a very large project, I used these two gentlemen as touch points. The first had already told the president and the minister of health about me, my mission, and my drivers. The second did so after I asked him to mention me, and in particular, my love for and general interest in their country. When I reached out to the president and the minister of health, I was quickly invited to visit. Six months later, I was running a major operation in their country, funded by the government, and treating hundreds of their citizens in clinical studies.

In another instance, I helped a Latin American country establish stem cell laws and translate those laws into a regulatory framework that guides their FDA. I did so for free and with no expectation of getting anything in return. I interacted with the president and the minister of the economy a great deal, and then went on to help one of them get connected to a world expert in a certain medical condition—again with no expectation of anything in return.

Years later, a billionaire whom I had never met, or even heard of, reached out to me and bluntly asked if he could invest in anything I was doing. Of course, I happily said yes! When I asked him how he came to offer me such support, he told me my name had come up during a meeting with the president, who was a good friend of his, and again during a meeting with the minister of the economy. Their conversation orbited around the energy I have for my mission, and he decided to follow that.

Over time, you will notice that employing the three-touch system steadily increases your network of supporters and manifests unexpected introductions and benefits. The invitation I received to meet with the Dalai Lama, though it was unexpected, came about because mutually trusted associates talked about me and my work. There's no way I can quantify the fun, productivity, and satisfaction I have experienced because of these people.

The three-touch system is not a single activity. It's more like a way of life—continually planting seeds and nurturing the soil to see what grows. That said, here are the SparkNotes:

- Maintain a large group of contacts who know and respect you
- Be accessible to others
- Take action
- Offer help with no expectation of return or repayment

I make myself accessible to everyone I come into contact with, and I help them for free. I take action. I offer my connections. I collaborate.

So, get out there! Share your resources; share your contacts; and watch how your world changes. Incidentally, employing this system makes you a better person— *ka-ching*, good karma!

NOW, TO BRING IT ALL HOME...

The concept of putting the right thing in front, regardless of company or industry, means identifying the *boat*...the one thing

130

from which everything follows in its wake. This one thing drives the entire mission. It is never ego, personal gain, quirks, or your baggage. While there are certainly professionals all over the world who bring this junk into the workplace, a mission never possesses such things.

The driver of a mission may be technological excellence, medical results, or methods of mass appeal; in other words, the driver is whatever X-factor is only obtainable through the convergence of the efforts of many people. If a CEO puts an unrealistic driver, such as personal gain, in front, the mission will inevitably fail.

You want your boat up front, not floundering in the wake of infatuation.

Unfortunately, I've noticed that businesses rarely put the right thing in front, which explains why 65 percent of businesses fail.[2] In biotech, that statistic is more like nine out of ten.[3] The reasons vary from company to company, but at the core, it's almost

---

[2] Deane, Michael T. "Top 6 Reasons New Businesses Fail." Investopedia. Investopedia, October 25, 2021. https://www.investopedia.com/financial-edge/1010/top-6-reasons-new-businesses-fail.aspx.
[3] Simpson, Stephen D. "A Biotech Sector Primer." Investopedia. Investopedia, September 21, 2021. https://

always because they put the wrong thing in front.

Every job requires a tribe, and no man, woman, or business is an island. You'll never get the good people to buy in if you're leading with something silly.

www.investopedia.com/articles/fundamental-analysis/11/primer-on-biotech-sector.asp.

7—KNOW THE GAME

Do you like macaroni? Because I don't—not anymore.

I used to. In fact, during the years I attended graduate school, had you been a fly on the wall in my shoddy apartment, you might have noticed an unusually large collection of boxed macaroni and cheese. To the uninitiated, it was a carb loader's dream: boxes upon boxes of microwaveable sustenance. While it was not nutritious (nor particularly delicious), it was my meal three times a day, for years on end. For those in the know, this was just life below the poverty line.

Each box of macaroni and cheese was sixty cents.

It was all I could afford to eat.

IT'S THE PUNCH YOU DON'T SEE COMING THAT KNOCKS YOU DOWN

There was never a period of time, in all my years as a student, when I wasn't working at least one job. I'm not a masochist, and I'm also not overly fond of hard labor, but the grants I received were never enough to pay for everything. Especially in the early years, it was a constant struggle just to subsist. Thankfully, I no longer have to hoard mac and cheese to get by. Yet, I think it's a situation every graduate student experiences to some degree.

It's the game. And knowing the game...and the cost of macaroni and cheese...allowed me to survive.

Knowing the game is just as important to an entrepreneur or a CEO as it is to a young graduate student. The hard truth is that if you fail to comprehend the rules of whatever matrix you're trying to navigate, the big dogs up top will not allow you to play.

Academia is no exception.

The political infighting and rigid structure of the university game can be summarized in one word: administration. I thought this was something I understood well, and

134

when I became a professor at University of California, Irvine, I eagerly jumped into it. In my naïveté, I thought to upset the balance without learning the rules. Before I knew it, my feet were bound, and I was up to my neck in committees and administrative distractions. If I got even one hour of my day to focus on the research I had come to do, it was a small miracle. Turns out, I had completely misjudged this particular game.

Disgruntled and discouraged, my inner narrative kept harping on the obvious: this wasn't the job I had signed up for, and it certainly didn't align with my personal mission. Ask any seasoned veteran of combat sports and they'll tell you—it's the shots you don't see coming that hurt the most.

And boy, were my eyes closed.

THEY'LL HATE YOU...UNTIL THEY RESPECT YOU

I had given it my all and the academic administration had chewed me up. I hadn't understood the game I had joined. At least not yet.

In an earlier chapter, I relayed that eureka moment when I decided to study this particular game, and then strategically rearrange it. "Wait a second," I thought, "I competed with 2,000 other people for this job, and they gave it to *me*! If I've resolved to quit, I might as well try to reshape this game board to fit my own vision of how it should be played." My naïveté and struggle were gone. I understood how the game worked now, and I just wasn't interested in playing it their way.

My principles were being negatively impacted by the weight of this adminis-trative load, so I shrugged most of it off. In retrospect, I realize I've always been something of a lateral thinker, and I sus-pect most other entrepreneurs are too. Why color inside the lines when there's so much untapped white space to work with?! I made my lists of what tasks I liked and disliked. Like some proverbial samurai, I cut ties with everything that was holding me back from my personal mission. I have to say, it was a fascinating psychological exercise (hello, innocent intrigue!) to watch the reactions of those around me. Some were very evidently appalled; others didn't seem to know what to make of me. Each and every one of them, however, had subscribed to the dogma of the game: *this is how things are done here.*

It never occurred to them to ask *why.*

It's not that they didn't agree with what I was trying to do. That wasn't really the issue. The issue was that I was threatening their system, their way of life. You see, the game protects those who have learned to play by its rules.

Innovation is almost invariably met with hostility—at least, at first.

When you are in a situation like that, you have to play to that sliver of common ground. In this instance, it was that fundamental desire every scientist has to help humanity, and I put that at odds with the game we were working within. I stressed I was not trying to dismantle the system, I just didn't want to play that way. It didn't work for me, and I refused to change my priorities.

It took about two years, but people eventually came around. Most of these chairpersons, even my most vehement opposition, ended up congratulating me for my actions. "You know," they said, "That was such an odd and enviable thing to do! Although we had to adjust to the extra workload, we also secretly emulated your actions in our own ways." In fact, in a university the size of UCI, there were thousands of other professors. I knew I wasn't burdening my peers with more work, and I had hoped that they would take some similar steps to improve their own workplace.

The academic game certainly doesn't like disruptors.

I made the critical mistake of not understanding the arena I was walking into when I accepted the position, and it almost made me quit, then almost got me fired from my dream job. But I was willing to put it all on the line to put the right thing in front—my desire to better humanity.

Knowing the game is *not* the same thing as playing the game.

LEARN THE RULES LIKE A PRO SO YOU CAN BREAK THEM LIKE AN ARTIST

Barry was an important mentor to me, though, if I'm honest, I'm not sure he ever knew that. He was a brilliant professor who guided me through four or five turning points in my career. If I was able to get away with anything in academia, it was because Barry had already busted down the door.

Needless to say, we got along famously.

Like me, Barry was an entrepreneur in a professor's shoes. He was also a very successful professor—lots of grants, lots of students, a big lab, doing very meaningful work. He started forming companies—he had three or four of them—and, before long, found himself under investigation for conflict of interest. It happens to every single entrepreneurial professor; Barry was just one of the first to do it, and a heck of a lot more brazen about doing it.

Looking back, I see he didn't really appreciate the value of nuance...to put it bluntly, the game kicked his ass.

He either didn't appreciate or didn't care that a university was filled with a bunch of administrators who were paid big salaries with excellent retirement benefits. Essentially, this amounted to the same thing as dealing with government employees; they can't lose their paychecks, they can't get fired, and they are religious followers of the rules—pretty much the diametric opposite of an entrepreneur.

Barry, on the other hand, was all entrepreneur all the time. When asked about the legalities of a conflict of interest, he just raised his voice and said, "I don't care about your interpretation of my work. You don't understand, and I'm not stopping."

Finally, one fateful morning, he met the campus police barring the entrance to his laboratory. The message was clear: "You are no longer allowed access to your lab." The chancellor of the university had a letter for him that plainly stated, "You may no longer utilize university resources." This was a fully funded professor under the scrutiny of an uninformed and heavily biased committee that had appointed itself the righteous defender of liberty. Again and again, they asked him, "Are you using the university resources to better your company?"

He wasn't doing anything of the sort, but because he didn't know the game, it blew up in his face. He didn't care to understand the underlying motivator of these administrators, to educate them, patiently prove to them he wasn't suborning any rules. He got truculent, and they said, "Well, this whole issue is about using university resources. Because you either can't or refuse to prove to us that you aren't using university resources, you're out."

Barry didn't back down. Instead, he set up a full laboratory in a biotech research park. He still had all his grant funding, so he used the money that would have gone to the university to help offset the operating costs of the new facility. He worked in a private setting doing his public work...the university's work.

I really admire him for that.

Barry told me a story once about one of his last and most fateful meetings with the conflict of interest committee. They were pushing on something, so he just got up and walked out saying, "Okay, let's reconvene next week." When he came back the following week, he brought a lawyer with him. He figured that since the administrators were talking about following the law and the regulations of the University of California system, it made sense to bring a lawyer into the conversation.

Bad move.

"That was a big mistake," Barry told me, "Because I acted like an entrepreneur." Barry was in the right, and the lawyer backed him up; but because he didn't know the game, his move didn't work. These administrators have limited rules they are required to follow, with a tremendous amount of gray area in policy documents pertaining to outside activity. They have the authority to interpret rules, and they are the ones who have to rubber-stamp things. In the end, despite Barry being in the right and having a legal expert backing him, they exercised their power not to "stamp" his documents.

He was out.

I learned a lot from Barry. Years later, I, too, found myself in a similar situation with an angry conflict of interest committee telling me, "This is not university policy." Instead of getting a lawyer, though, I turned on a computer and said, "Well actually, here's the university policy right here. Let's take a look together. Help me understand." I let them read and interpret the document to me in a terribly patronizing manner; but as they did so, I could feel them begin to bend their conclusions to conform to the very policy they were referencing. One painful hour later, they backed down. I was right, and they were wrong; but instead of challenging them, I educated them. I didn't get kicked out, and I didn't lose my laboratory. I owe that to Barry. Through his experience, I learned the game of career administrators.

Unfortunately, that wasn't the end of things. Ultimately, that initial meeting became part of a longer effort by some disgruntled professors, who formed a "Committee for the Management of Hans Keirstead," sanctioned by and reporting to a senior research administrator. This group of nine professors met every month to analyze everything I had done in my company, and in my lab, during the previous thirty days.

It might have been flattering if it wasn't so downright vexing.

After a year, it was determined I had a manageable conflict of interest, and the committee was disbanded. To this day, I have no idea what "manageable conflict of interest" means, but I took a lesson from Barry, took my win, and chose not to press it.

Four months later, I received a cease and desist letter from a senior research administrator telling me to stop all outside activity. "Here we go again," I thought. I reminded the administrator that I had been cleared by the conflict of interest committee, and responded to him with a flat-out no, as I had no conflict of interest and was abiding by all university system rules. His response was to inform me that a new committee called "The Perception of Conflict of Interest Committee" had been formed by several of the former conflict of interest committee members on their own initiative, and that they had made this adverse ruling.

"The Perception of Conflict of Interest Committee." Wow. I will never cease to be amazed by the machinations of university administration.

I never did abandon my outside activities, but I did agree to meet with this new not-so-perceptive committee of nice, full professors every month for a year to discuss their perceptions.

A worthwhile use of funds and time, I must say.

I admit I didn't completely abandon Barry's approach. At one point, while in front of the committee, I volunteered to bring in my lawyers to educate the committee on the university system rules. I wanted to prove to them beyond a shadow of a doubt that I was in full compliance with the law.

The committee responded by freezing all of my grant funds for three months while they investigated further. I had 48 people in my lab at the time, with multiple grants totaling about $3 million per year in funding. I chose to pay the rent and mortgages personally for eight of my staff during that time. Three months later, after nothing further was ever requested of me and no meetings were held, I was informed that no indiscretions had been found, that there were no improprieties on my part.

One member of that committee told me, "Never mention lawyers again, okay?" Those three months were my penalty for not respecting the game.

Learn the rules like a pro so you can break them like an artist—or an entrepreneur.

THE INTERVIEW THAT CHANGED EVERYTHING

Whether writing grant applications or pitching to a room of potential investors, my whole career, both inside the university and out in the real world as CEO, has had a very large fundraising component.

Yes, there is a fundraising game, and you'd better believe you have to learn its rules.

When I was approached by *60 Minutes* to see if I would have an interest in a tele-vised interview, the senior administration at my university strongly encouraged me not to accept. They were worried that, even though it would bring tremendous exposure to my work at the university, the potential backlash would be catastrophic if it went poorly. They offered a lot of valid examples where *60 Minutes* had cast interviewees in a negative light. I'll admit, they had a point.

But you have to know the game you're playing, and for me, this was a fundraising opportunity like no other. The benefit of exposure outweighed the risk I would be vilified on camera. I decided to go for it.

Spoiler alert: it turned out great.

In all honesty, I couldn't have asked for a better yield from one interview. That particular episode was a full one-hour show focused on my stem cell work, which they aired several times. As if I wasn't tremendously grateful for that in and of itself, American Airlines showed the episode on every flight for a year!

Needless to say, it garnered tremendous exposure.

It was so broadly televised, in fact, that the founder of Pacific Investment Management Company (PIMCO), Bill Gross, reached out and introduced himself and his wife Sue. At the time, Bill Gross was one of the wealthiest men in the world. He was also voted the second most influential man in the

U.S., trailing behind the chairman of the Federal Fricking Reserve! Whatever he saw in that *60 Minutes* piece must have struck a chord; soon after he wrote me a check for $10 million to be used for whatever area of research I saw fit. In their honor, I created the Bill and Susan Gross Stem Cell Research Center, the facility I directed for its first five years.

That *60 Minutes* interview was truly the gift that kept on giving. We gained a tremendous amount of credibility, and that led to donations—somewhere near $100 million when it was all said and done.

Another unexpected but amusing offshoot of this sudden publicity was my *Men's Vogue* magazine photoshoot—what, you didn't know scientists could be sexy too?

I took a risk, but it was a calculated risk, and it paid off in ways I couldn't have predicted. That one powerful piece of tele-journalism produced quantifiable results that included a full research center and millions of dollars in direct donations to my work.

For a hot minute, it was all sunshine and roses.

The thing about rising into the spotlight, though, is you inevitably put a target on your back. In that regard, I was no exception. As an entrepreneurial scientist and academic, I've experienced what happens when the various games I need to know collide with one another. Carving my own path as an entrepreneur *and* a scientist, I fought battles on both ends of the spectrum: against venture capitalists who said I wasn't committed enough to the business because of my university work and against my colleagues at the university who viewed me as some sort of money-motivated alien.

For the most part, I traversed these two worlds fairly well. At one point, I was running a young company valued at around $100 million and had successfully obtained federal grants for my science totaling over $3 million a year. Additionally, I had the research center to manage and a batch of employees who reported to me. But things started to get wonky after the *60 Minutes* piece...

Slowly but surely, my research papers were receiving a higher percentage of rejection letters. At first, I couldn't make sense of it—this was some of the best, most airtight science I'd ever produced and it was consistently being turned down. And then I noticed my grant application scores were being downgraded as well, making it harder to get grant money. Things finally came to a head when an article about my work was published in the *L.A. Times*. The catch line of the whole piece was essentially: if someone's going to cure spinal cord injuries, it's going to be this guy.

Flattering, right? When I started reading, I was thrilled and grateful to get such detailed coverage of my work. As I read on, though, the article became extremely critical. The writer had interviewed nine world-famous spinal cord specialists, every single one of whom was at least twice my age. Each of them gave some version of the same narrative: Keirstead is overreaching. He's saying this method of making rats walk again will be used to treat paralyzed humans. It is unlikely this will ever get to humans. And I think he's playing out false hope for the community.

All nine of these famous scientists had said the same thing.

This is the point where *shut up and do it* meets *know the game.*

One year later, my treatment was being applied in hospitals and was proven to regenerate *human* spinal cords.

Since then, every single one of those professors has invited me to be a keynote speaker at a conference or some other award ceremony. And so, I learned the last lesson of the game: the brighter you shine, the easier a target you make.

To this day, I remain extremely dismayed by the pettiness of some colleagues. After all, weren't we all in this together? Wasn't this supposed to be about science, about the betterment of society at large? But as disappointed as I was with my colleagues, I did frame that *L.A. Times* article. Visitors to my office often ask why I display an article filled with such skepticism from my peers. I only smile and think of Timothy Atchison, the first quadriplegic patient who regained the use of his arms and hands after receiving my treatment. I haven't always figured the game out in time, but I did get that right.

EXPLAIN, EXPLAIN, EXPLAIN

Over the years I've realized that acting in isolation, as certain as I may be with my direction, attracts the negative, the doubters, the jealous, and the uninformed. Taking the time to talk with people about exactly the kind of work you're doing, and more importantly, *why* you're doing it, can have an incredibly alchemizing effect in changing their perceptions and motivations.

In the absence of knowledge, fantasies evolve.

As a scientist, it's imperative to me that other scientists, investors, and patients know what I'm putting in front—that my mission comes from a desire to better the world. It is not a spotlight-on-Hans campaign. It's a shut-up-and-do-it campaign. It's leading with balance. It's putting the right thing in front. It's everything I've learned along the way that's shaped me into who I am.

So, while it sometimes feels like overkill, I've learned that playing this particular part of the game requires superfluous explanation. It means being absolutely certain I'm telling my story from a place of authenticity, empathy, and sincerity. The onus of this entrepreneur's burden is yours and yours alone.

But once you know the game.... anything is possible.

8—LEADING IN A BUGGY WORLD

F orget your zombie apocalypse or nuclear world war—we've got bigger problems.

Every biologist in the world understands the most prevalent threat to humankind comes from bacteria and viruses—commonly referred to as bugs. As the icecaps continue to melt and genetic mutagens continue to increase, our species is being exposed to more and more varieties of bugs for which we have no immune tolerance. Moreover, these bugs are not some tired buffet line of old bacteria; they constantly evolve, adapt, and change.

THE END OF THE WORLD STARTS WITH A COUGH

Here's a bit of underrated pub trivia for you: did you know every drop of water on our planet came from outer space? That's right, all the water in your bathtub, sink, cooler, ice chest.... all of it hitched a one-way joyride to Earth through space. Thank you, meteorites. The downside of that intergalactic carpool was that it didn't just bring droplets of water and space rock; it also brought foreign microbials which have evolved into the bacteria and viruses that keep us bowing our heads to the porcelain gods, year after year.

Needless to say, it's a buggy planet, and our ability to avoid these naïve geniuses is extremely limited. They have developed over millennia, and it's not without a certain reverence I say they are very, very good at what they do. In fact, five to eight percent of our human genome is viral—that's how good these bugs are at invading humans and going undetected! They've quite literally knit themselves into our DNA. At the microscopic level, they get into our bodies undetected where they flourish and propagate, storing themselves immortally in our biomes—indeed, into our genetic makeup as a race.

Disturbing *and* extraordinary.

The greatest drug ever developed was penicillin. The world and its human population would look a whole lot different today if penicillin did not exist. For one thing, there'd be noticeably fewer of us taking up real estate on this Little Blue Dot. While penicillin does excel at destroying bacteria, we know now that's only half of the deadly equation. There is currently no analog that effectively neutralizes viruses. Each new iteration that comes along requires a vaccine that works specifically on that virus in order to protect us (e.g., your annual flu shot). Influenza is a virus, and every year the vaccine you receive has been specifically designed to address the version of that virus *currently in circulation.*

Unfortunately, neither penicillin nor any other antibiotic effectively combats viruses.

The scientific community is so concerned with viruses, in fact, that there are labs around the world dedicated to monitoring and tracking the existence of emerging or developing strains. When something potentially dangerous pops up on the radar, the worldwide alarm is sounded. This process occurred with the H1N1 and Ebola viruses, and most recently, with the SARS-CoV-2 or COVID-19 virus.

SOUNDING THE ALARM

I first became aware of COVID-19 in November 2019.

A doctor in a lab in Wuhan, China, sounded the international alarm about a new virus, and I immediately started paying attention. As a scientist and biologist, I had a decent understanding of the history of viruses, and a founded fear of the havoc they can wreak on humanity.

However, as November progressed into February 2020, the world's attention shifted from China to Italy as the epidemic took hold there in full force. Italian medical professionals from up and down the boot began to set aside the international medical norms of lengthy peer review to get their views across and took to public international forums and the media to warn us: this was going to be much, much worse than we had anticipated. It was around that time I realized this virus was going to have far-reaching impacts.

It was time to mobilize the team.

The first action I took was to assign myself and two doctors—my chief science officer and my chief medical officer—a task. The three of us were to monitor and study the movement of the virus continuously. Each Monday morning, we would meet to discuss what we had learned. I also welcomed the leaders of each of my company divisions to join these meetings.

In an effort to be as efficient and transparent as possible, the meetings followed a strict protocol. I would be the first to lead the discussion, offering my insights and opinions regarding what I had learned over the previous week. Next up was my chief medical officer. The rule was this: He would not repeat anything I had already mentioned. He would either start with "I agree with everything Hans has said" or "I disagree with one (or more) of Hans' conclusions, and here's why." He would then add his own insights and opinions. Next to speak was our chief science officer, who followed the same protocol. I developed this protocol after a casual discussion with a friend of mine who had won the Nobel Prize. While discussing the new circles he walked in, he mentioned one of the greatest benefits was the efficiency of communication amongst intelligent people. He relayed a story of a two-hour meeting with some of the best minds on our planet, remarking that not one person repeated themselves for emphasis, no one raised their voice to control the conversation, and no one reiterated what another had said. Everyone was intelligent enough to remember all that had been said, and at the end, the group made a decision that took into account every angle that had been proposed.

Respect the insights and opinions of your colleagues. Keep a sense of innocent intrigue so you can value a new or opposing view. Leave your ego at the door.

The final chapter of each of my weekly company meetings addressed the big follow-up questions: given all of this context, should our company policy change, and should our research approach change? I would suggest an answer for each of these two questions, and every leader would then verbally give their consent or dissent. The result was that in just under thirty minutes, the company's top three scientists had updated a large portion of the management team.

Soon thereafter, I implemented a work-from-home policy. To the best of my knowledge, we were one of the first companies to do so—this was back in early February 2020. We coached the staff about personal protection and behavior and also outfitted every room in the company with masks, gloves, and HEPA filtration devices to filter the air. I also tasked one of my doctors to source the best testing kits available in the world.

Not all companies are, or will be, in a position to contribute to a solution during a national or international crisis. My biotechnology company, however, was equipped to develop treatments for serious diseases and, at that time, was testing a cancer vaccine in 16 hospitals across the United States. We were in the right place to take action, but it wasn't just my decision to make.

In March 2020, I called a meeting of our department heads and we determined we had the capacity, resources, and know-how to generate a solution: My team was on board. Immediately after, we began work on a COVID-19 vaccine. I'll never forget the faces of my staff as they began their tireless journey to fight this pandemic.

WALKING THE WALK, TALKING THE TALK

All of these actions reflect the motifs I've talked about thus far in the book—innocent intrigue, people, balance, putting the right thing in front, etc. In stepping up to create a vaccine, we were about to become reacquainted with my old nemesis: the game.

If you haven't guessed by now, there is indeed a vaccine game. This particular game comprises two parts: the approaches to creating vaccines and the regulatory process for getting a biologic approved. And that game is set within a dynamic framework of extreme politics which always accompanies the build of multi-billion-dollar industries...that framework being different in every country in the world.

In all approaches to creating a vaccine, the ultimate goal is to get a class of your immune cells, called *antigen-presenting cells*, to take up and express parts of the virus, called *antigens*. These antigen-presenting cells show the viral antigens to a host of other cells in the immune system, triggering a massive readiness response to the virus. In essence, the cells teach the immune system to be ready for the real invader: "Here's what the intruder looks like, suit up, and go take it down!" The antigen-presenting cell is like the orchestra conductor of the symphony that is your immune system; it is the general who controls every fighting force in your immune system.

Every scientist in the world working on a COVID-19 vaccine is trying to get your antigen-presenting cells to take up, read, and express viral antigens. The goal is the same regardless of nationality, age, or political structure. What's different, however, are the methods for accomplishing this.

All the existing vaccines load your blood with viral antigens, hoping some small percentage of those viral antigens are taken up by your antigen-presenting cells. These vaccines load your blood with viral antigens by genetically modifying your own muscle cells to over-express particular viral antigens, or they directly load your blood with viruses that make huge amounts of viral antigens. Those excess free-floating viral antigens are responsible for most of the side-effects of vaccines—the headaches, chills, nausea, and fever. But they do work, supplying your antigen-presenting cells with viral antigens to induce a memory of the bug and, hence, an immune mobilization plan for defense should you ever experience infection.

By now it should come as no surprise that Team Keirstead usually takes a lateral approach; I seem to have modeled my career on it. By leveraging similar methods used in our cancer treatment, we draw blood from the patient, isolate their own antigen-presenting cells in a dish, and train them to take up, process, and express the SARS-CoV-2 antigens. From there, we can inject the personalized vaccine

161

straight back into the patient. There are no free-floating viral antigens in our injection, only the patient's own antigen-presenting cells that were educated in a dish. It's a fully personalized approach because we are using the patient's own immune cells rather than a generalized vaccine. This was the first time in the world such an approach had been developed to vaccinate against a pathogen.

This approach was incredibly exciting because it would help (or so we believed) in cutting through some of the FDA red tape. A major regulatory accelerator for our COVID-19 vaccine would be the ability to cross-file it against our cancer vaccine, which had already gone through the lengthy FDA approval process and had been approved for Phase 2 and Phase 3 clinical testing on cancer patients. Both vaccines use the frontline warrior of the immune system: the antigen-presenting cell. In the case of cancer, we load the cells with antigens from tumor-initiating cells we isolate from resected tumors. Every cancer patient has unique tumor-initiating cells with hundreds to thousands of unique antigens. With COVID-19, we simply replaced the antigen load, once again with unique antigens, this time from the SARS-CoV-2 spike 1 and spike 2 proteins.

We've proven this procedure to be extremely safe and efficacious in hundreds of people with cancer, so a cross-filing would assure the FDA that the safety, manufacturing, and quality systems from the cancer vaccine apply to the COVID-19 vaccination. This would allow us to skip the preclinical animal experimentation required for new drugs and move straight into a combined Phase 1-2 clinical trial. This would save years of research, millions of dollars, and the lives of countless animals, and get a safe and desperately needed vaccine to humans faster. A plan for the drug was emerging!

Building a safe and efficacious drug is one thing. But ensuring widespread access to that drug is another thing entirely. And let's add one more contextual challenge: How do we prevent economic decline in those countries that must continually buy vaccines and distribute them at super cold temperatures in-country? What vaccine strategy can address all of these things?

The currently available vaccines are all made in multi-billion-dollar centralized manufacturing facilities in whatever country the vaccine producer is domiciled. If some other country wants to buy vaccines, they face the harsh reality that billions of dollars will leave their economy every single year, for the foreseeable future. Billions leaving Indonesia every year to China. Billions leaving African nations every year to Britain. This economic vaccine bleed is clearly impossible for emerging nations to undertake or sustain. When the United States had vaccinated 49 percent of their population,[4] 130 nations in the world had yet to receive a single COVID-19 vaccination.[5]

So why not build a system whereby COVID-19 vaccines could be produced in *any* country using materials sourced primarily from that country? Limitless distributed manufacturing.

---

[4] *Suliman, Adela, Bryan Pietsch, Brittany Shammas, and Amy Goldstein. "Covid-19 Latest Updates: U.S. Hits Biden's Vaccination Goal a Month Late, with 70 Percent of Adults Receiving at Least One Shot." The Washington Post. WP Company, August 3, 2021. https://www.washingtonpost.com/nation/2021/08/02/coronavirus-covid-live-updates-us/.*

[5] *Andrew, Scottie. "More than 130 Countries Haven't Received a Single Covid-19 Vaccine, While 10 Countries Have Administered 75% of All Vaccines, the UN Says." CNN. Cable News Network, February 18, 2021. https://www.cnn.com/2021/02/18/world/united-nations-130-countries-no-vaccine-trnd/index.html.*

Flip on that innocent intrigue. Flip off the naysayers. Find your points of leverage, put the right thing in front, then just do it.

The right thing to put in front of this one was most certainly a best-in-class vaccine. We hit that hard, and we nailed it. And what was my best leverage for the task at hand? A connection to the most senior leaders of Indonesia and a pharmaceutical company there, who—much to my delight—agreed to provide all the funding to test our vaccine in Indonesia.

To address that economic decline thing, we designed a kit rather than a vaccine. A kit that could be assembled in any country by a local kit assembly plant, sourcing as many of its components as possible from within that very same country. That kit could then be distributed to health centers by distribution companies within that country. And vaccines could be made using that kit by health care professionals within that country. Thus, the vast majority of the government's cash outlay would remain within the country, effectively creating an economic stimulus while vaccinating its populace. The greater plan was set!

Entrepreneurs beware: innovation is always suppressed by convention.

While we nailed the biologic, having created a safe and efficacious vaccine, the game kicked us around like that little metal ball in a pinball machine. The feedback from the regulatory agencies was a litany of critiques that pertained to other vaccines but had nothing at all to do with the biology of our vaccine.

The feedback from the government, industry, and financial partners was a barrage of excitement and confusion over a plan with which no one in the world had any experience. We were bucking the system. We weren't using the usual approaches that established vaccine companies used. Seasoned veterans in vaccine development, biological product marketing, reimbursement, supply chain, product scale-up... were all rendered newbies in a system that did not require their vast experience. It was like hearing all those voices from my past saying, "Oh no, that's not the way it's done! That's not the way things work! You do not run a $300 billion company! You're over-reaching." It was like those irrelevant academic committees trying to fit my square head into their perfectly round policies. We knew our approach and logic were sound, but these are the administrators, and nothing constructive can be accomplished by arguing with them.

The situation called for balance, for myself as a leader, and for my team. We called everybody together, this team of highly dedicated individuals, and we made a list of every challenge. I have to say, it was quite an extensive, impressive, and intimidating list! But then we put one thing in front—the excellence of the medicine—and realized every one of the myriad challenges would fall into place behind that lead, the one thing from which everything follows in its wake. Focus on that, get excited by that, and you will be amazed at the support group that grows around you to take care of the myriad issues your focus manifests in its wake.

I can't tell you how many diverse questions I answered with the single phrase: "Well, we have a vaccine that is safe and works, and can be made in any country. Let's work with that." And they did. The Indonesian regulatory authorities eventually understood our approach and approved our clinical trials. The military suddenly invited us to use their phenomenal hospitals, equipment, and impressively trained staff. The government signed an economic decree providing support for all clinical activities. Even the parliament stepped in to facilitate patient recruitment and approvals.

I hope our vaccine is eventually commercialized but, for the purposes of this book, it is the approach, perspective, and journey that are important. A few invaluable lessons have been reinforced this year that I think every CEO could benefit from hearing:

- **Have a readiness plan**. It is best to assume we will have to face COVID-19 viral mutations or other pandemics in the future. Companies need a readiness plan for such things, with policies and procedures clearly spelled out so leaders can move quickly to put critical operations in place.
- **Keep a cash reserve.** Whatever you have now, add to it. One of the most common reasons for the failure of companies in times of emergency is insufficient funds. I was advised by a top-notch venture capitalist some time ago always to raise more money than I think I need. When I resisted the idea, citing fiduciary responsibility and the risk of dilution, he said, "Almost every company that fails would not have done so if they'd only maintained adequate cash reserves. You need a bank account for a rainy day." I realized he was right. As far as dilution goes, suffer a little dilution now to gain the confidence of

having a cash reserve to use in an emergency.

- **Treat your people right.** This is not simply about dealing with an emergency. This is a "people are everything" item that applies specifically to your team. Here's the thought process: if you consistently and unerringly demonstrate your respect and consideration for the well-being of your employees, they will not panic or scurry during difficult times. Treat your people right, and they will follow you to the end of the world —and all the way back out.

WHAT'S NEXT?

Nothing is going back to "normal."

This COVID-19 bug will always be with us, culling a significant percentage of our population year after year. Herd immunity will never be reached on a local scale given the large percentage of people who do not want a vaccine, and it will never be reached on a global scale given the large percentage of poor people who cannot gain access to a vaccine.

The COVID-19 pandemic will leave lasting changes in the business world. We have learned that face-to-face meetings are not the necessity they once were, nor will all employees be required to be in a common workplace for large portions of the day. Video conferencing and online team productivity tools were initially adopted out of necessity, but now leaders all over the world are realizing there are major time and monetary incentives to be gleaned here. We are entering a new era of what will be considered "the new normal."

The new ways of doing business will change the world. Airlines have been dramatically affected by this pandemic. There is talk of returning to a system that requires international travelers to show both proof of immunization and a passport for domestic travel and globe-trotting. The emergence and spread of SARS-CoV-2 is certainly related to air travel, but it is also related to urbanization, habitat destruction, intensive farming practices and live animal trade...all of which will forever be changed by the appearance and spread of this bug. The dissemination of information on infection agents will also be forever changed, as the world has experienced once again that infection travels faster than policies can be made and instituted. We have seen an erosion of trust in governance, widening geographic inequalities with respect to health care, and the widening socioeconomic inequalities with respect to the impact of disease on a person or family. Labor markets and the workplace are forever disrupted. I, along with a group of business and government leaders, am involved in a very real way in figuring out how we can avoid these types of situations again in the future.

A lasting change that is close to home for me, and one I am very grateful for, is a newfound respect for science and medicine. It's a simple fact that you may love your house, your car, your lifestyle, and your bank accounts—but all of those things go to dust in the face of a personal health crisis.

From the ashes of tragedy, we have an educated populace that respects science and medicine now more than ever. The majority of our populous now understands what a vaccine is, they understand something of the process of getting a vaccine approved, and they understand the billions of dollars that vaccine development entails. This general education is invaluable. This knowledge helps in making the right vote for NIH funding of medical support and community development, where ignorance in that area has led to the indiscriminate decimation of research, the careers of brilliant researchers, scientific communities, and ideas alike. It helps elect the right public servants up and down the line. It helps us listen to the scientists and the medical professionals actively fighting to save our lives.

Everything is science.

We need more government representation, but we just don't have it because science is often unapproachable by the voting lay public, often incomprehensible, and it takes place in ivory towers. Congress has a dearth of medical and scientific advice. There is lack of regulation in some areas, antiquated regulations in others, while still others are overregulated. Most regulations are driven by Big Pharma lobbying, and they sometimes have excellent outcomes. Other times, it's plain to see the destructive path we are set on, as money is valued over health.

Vocal scientists are the answer.

Scientists need to stand up and take a podium from time to time. I speak to high school students, and I speak to Congress and lawmaking bodies all over the world because I want to educate people that science is something we need to bring into the mainstream.

We need people like Carl Sagan and Bill Nye and Stephen Hawking.

There are myriad young, inquisitive, and naturally eager minds out there. Minds that are brimming with the possibility of discovery. But science is so under-funded that people leave as quickly as they come. Students look at their professors working eighty hours a week while struggling financially; they notice the disappointment of not getting the accolades and recognition they deserve. They see the bitter fights and competition for papers, publications, and grants. And because there's so little money in the system, it becomes furiously competitive.

Science is hard, and *the game* is harder.

If we can't put science front and center across the world, the bugs are going to wipe us out. Worse yet, human genetic engineering—without proper regulatory channels—will do us in. We're just not scienced up enough. The hard truth is that we've got to do a better job of stepping up and speaking out.

LAST WORDS

My path has never been laid out before me. I've always hacked my way through a jungle with a machete and a certain faith in the unknown. To be honest, I loved that. I've never been dejected. I've never thought to myself, "Why do *I* have to go through this?" I'd be lying if I said it wasn't difficult; there's a lot of damn jungle out there and just as many predators. People watching from the sidelines with a cold beer in a Koozie ask you, "How come you're not moving faster?"

But attitude is everything. If you love what you do and you don't take yourself too terribly seriously, there is the possibility, however miniscule or far-fetched it may seem, that you might just change the world.

So, trust the compass, not the map. The latter changes. You shouldn't.

I'll leave you with this recap of what we've covered:

- **Innocent Intrigue**—When was the last time you realized something major? The last time you had an "Aha!" moment that rocked your world? That was innocent intrigue. The pure, unadulterated liquid of exploration. We know this inherently as children, but as we grow up, our sometimes-jaded minds begin to believe that the magic in the world has somehow been greedily siphoned out. There's nothing left to see or to explore, or at least that's how we learn to act. As a scientist, I can tell you these moments of exploration are indescribably important. They keep the brain matter healthy, active, and agile—and those are the makings for great leaders. COVID-19 may not have been a very happy period of innocent intrigue, but it rocked our world violently out of its comfort zone. What we choose to do now, how we go forward from here, will define our leadership practices. You've been given the chance to see things from outside of your usual perspectives, so explore them! There is plenty out in the world to stun and awe you, but it's your responsibility to go out there and seek it.

- **Try, or Just Shut Up and Do It**— You know the best thing about growing up on a farm, besides the freedom, that is? It's quiet. There are no city buses rumbling by, no blaring horns and, most importantly, no one complaining loudly into a cell phone. The life of a farmer is backbreaking. It drains the fascia of your being from dawn to dusk until you slink back to the quiet safety of your bed and sleep the sleep of the dead. Now maybe that doesn't sound so appealing to you, but I wouldn't have traded my childhood for anything. It taught me one of the most valuable lessons I've ever learned: *just shut up and do it*. Whatever you set your mind to, so long as your intentions are worthy and just, do not be swayed by hardship, the number of obstacles, or the immensity of the challenge before you. Once you commit to an idea, do not waste excessive mental energy imagining how it will play out. Don't build your own barrier to entry. Just do it. Commit to that which calls you forward and wakes you up from pleasant dreams.

- **People are Everything**—If, during your professional life, you find yourself losing steam, or if you've tripped the autopilot button, or the omnipresent *why* driving your actions begins to fade into the hazy background, I ask you to do one very simple thing: look up. Look up from your screen, from your notes, from the jangled internal narrative playing forwards, backwards, and sideways through your mind. Unless you've become the first person to turn yourself into an island, and have created your business entirely on your own, then you should be face-to-face with your *why* factor. People. It's always been about the people. The moment your internal compass starts to waver, I encourage you to tap into your people. Remember that they lean on you for support. They are the web that keeps the structure together, the foundation that keeps it standing upright, the oil that lubricates the machine. Treat them with integrity and they will go to the ends of the earth with you.
- **Balance is the Other Everything**— Meditation, breathing techniques, yoga...these are not just the lofty pastimes of Tibetan monks and secluded

spiritual hermits. In fact, they have become more socially and culturally relevant than ever before. Never in history have our leaders faced so many distractions: email, text messages, board meetings, soccer practice, marketing calls, lectures, seminars, social media, etc. I could go on and on. Some of these distractions have bettered our lives, no two ways about it. However, the vast majority simply eat up our time, cause us stress, or both. We have fallen prey to mindlessness and inattentive uses of our cognitive capacities. Therefore, the CEO, the entrepreneur, and the leader must exercise daily the practice of shutting out distraction. Balance. That elusive, right on the tip of your tongue sensation. Balance in a leader is where innovation meets pragmatism, where tomorrow's results find their foothold in today's investments. Balance is the other everything.

- **Leading with Balance**—Visionaries are just productive people with OCD. Sure, they are the captains of the ship, the masters of the enterprise, but it takes an entire army to make one corporation successful. Visionaries are the driving force behind creative disruption. They are the game changers who come along and rile us up, the ones who challenge stagnant notions of the status quo. Yet, for all that, visionaries are only as good as their counterparts. The most effective CEOs *lead with balance*. What does that mean? Well, to put it bluntly, it means knowing when the hell to get out of your own way and let people do the jobs they were born to do. Just because you are the visionary does not mean you are the most important player on the field. A good leader knows his or her job is to galvanize the troops, not necessarily to command them. Put the right people in the right places and productivity happens. The paradox of leading with balance is that the leading part requires you to let go of some measure of control. Balance means trusting others to do the things you've assigned to them and giving others the dignity to prove themselves capable.

- **Put the Right Thing in Front—** Thought experiment: would you rather have a Porsche today and a foreclosure on your home one year from now, or a medium salary today and royalties, residuals, and dividends paid to you for the remainder of your natural life? Pretty easy, right? And yet, all too often, people fall for the bling. The "get rich now, worry about all the bad stuff later" scheme. It doesn't work like that. You can't reverse-engineer success by purchasing the overpriced sports car, a high-rise office, and an employee paintball arena. These things are the byproducts of years of laborious sacrifice and hard decisions. Leaders who set out to make their own path must know this from the start. You have to put the right things in front. You have to know what your priorities are and what you're willing to sacrifice now to get something later. Temptation is a cruel mistress, and she only gets more persistent the more successful you become. Remember your *why*, remember your people, lead them with balance, and keep all those things front and center as you swat away the beautiful but, ultimately empty, distractions.

- **Know the Game**—Ask any prize fighter in the world...you can condition your body to withstand remarkable damage with incremental training, but the second you lose concentration, the moment you forget where you are, that's when you're going down. It's the punch you don't see coming that knocks you out. Whatever arena you fight in, be it a corporate one or otherwise, you have to understand the rules of the game. If you don't know the ins and outs, the loopholes, and backdoors, you're opening yourself up to procedural mistakes, accidental occurrences, and a plethora of bureaucratic agencies I promise you do not want to mess with. From a strategic perspective, being a business leader is only slightly different from being a military general. Just as it would be grossly irresponsible for a general to send his troops onto a battlefield without knowing the terrain, it is equally deplorable to lead your employees into an industry in which you have left them all financially defenseless. Before you take your first step, hire your first employee, or write your first check, make sure you know the game you've signed up to play.

****

The whole really is greater than the sum of the parts.

Innocent intrigue, people, balance, knowing the game, putting the right things in front—they are all universal truths, truths that, like a compass, have guided me in life. Each theme influences the others, creating a path that is nothing short of life, the universe, and curing everything.

ACKNOWLEDGEMENTS

A central theme of this book, and indeed my life, is the motivation, inspiration and drive provided by the many incredible people whom I have encountered. I am humbled by you and indebted to you for providing the pivots that have shaped my every experience on this Little Blue Dot. I thank every person who inspired the stories in this book, every student and employee who inspired me to be more, and every mentor who provided the guideposts for my career and life.

When asked by a colleague what it was like to be married to me, my wife Niki replied, after a thoughtful moment, "It is like flying a kite in a hurricane." I will forever be thankful to this brilliant woman for putting up with me, for grounding me, and for coaching me. To the reader I advise: always marry someone smarter than you.

I thank my father for proving to me that anything is possible, and my mother for showing me that the way is always kindness. I thank His Holiness the Dalai Lama for welcoming me into his heart and home, and for gifting me with an eternal smile that will never fade. I am indebted to Deborah Bannon and the Leaders Press team for their commitment and efforts to publish this book. And I thank my friends and family for the million memories that I carry with me and reference, like Siri, on a daily basis.

ABOUT THE AUTHOR

Hans Keirstead, PhD, is an internationally known stem cell expert, professor, and serial entrepreneur who led the development of a treatment that restored movement and sensory function to people with quadriplegic spinal cord injury, and a treatment for cancer that has saved the lives of people with melanoma, brain cancer, and ovarian cancer. He has also led therapy development for immune disorders, motor neuron diseases, and retinal diseases, as well as a vaccine for COVID-19. He has founded four successful biotechnology companies, each returning significant value to investors and treatments to patients. He holds board positions in several prominent biotechnology companies.

Dr. Keirstead received his PhD in neurobiology from the University of British Columbia, winning the Cameron Award for Outstanding PhD Thesis in the Country.

He conducted postdoctoral studies through a fellowship at the University of Cambridge. He received the honor of election as Senate Member at the University of Cambridge and Fellow of the Governing Body at Downing College, the youngest member to have been elected to those positions.

Dr. Keirstead was Professor of Anatomy and Neurobiology at the University of California at Irvine. During his 15-year tenure, he was awarded the Distinguished Award for Research, the UCI Academic Senate's highest honor, as well as the UCI Innovation Award for innovative research leading to corporate and clinical development. Dr. Keirstead consistently raised over $1 million a year in peer-reviewed grant funding and, while Assistant Professor, founded the UCI Sue and Bill Gross Stem Cell Research Center, which he directed.

A notable figure in the field of regenerative medicine for over three decades, Dr. Keirstead has mentored over 100 students, published over 100 manuscripts, and been granted over 20 patents. He was a founding advisor of the California Stem Cell Initiative that resulted in a $3 billion stem cell fund at the California Institute for Regenerative Medicine (CIRM). He has been a long-time advisor to several governments on biomedical policy.

Dr. Keirstead's work earned him the distinction of being featured as one of the 100 top scientists of the year in *Discover Magazine*. He was featured on TV's *60 Minutes* in a full segment covering his treatments, and he and his research have appeared in *Newsweek*, *WIRED*, *Esquire*, *The New York Times*, *TIME Magazine*, *Men's Vogue*, *Science*, *The American Spectator*, and *NPR's Biotech Nation*, among others. He has given dozens of keynote addresses at industry meetings and events, as well as popular thought leadership conferences such as TEDx and SXSW, was honored as one of the seven top thought leaders in the nation at the Seven Wonderer's event by Beakerhead, and was named one of the "Top 20 in 2020" by the Consulate General of Canada for his contributions to COVID-19 research.

Among a long list of personal accomplishments, Dr. Keirstead has a black belt in tae kwon do and is a helicopter pilot.